By the Numbers

By the Numbers

Numeracy, Religion, and the Quantitative Transformation of Early Modern England

JESSICA MARIE OTIS

OXFORD
UNIVERSITY PRESS

To Mom and Dad

Contents

Acknowledgments

As a sophomore in college, my academic advisor in the mathematics department sent me trotting across the campus of the College of William and Mary to declare a history minor. Little did I know that those were the first steps on a journey that would lead me to a PhD in history and this book. A truly humbling number of people have helped me get from there to here and I am incredibly grateful for their support.

I want to begin by thanking the librarians and archivists who made my research possible. The staff at the Huntington Library and Folger Shakespeare Library were especially welcoming and provided homes away from home for me at various points while I was working on this book. Thanks also to the staff at the American Philosophical Society Library; British Library; Bodleian Library at the University of Oxford; Cambridge University Library; Carnegie Mellon University Libraries; Essex Records Office; George Mason University Libraries; Guildhall Library; National Archives of the UK at Kew; Lambeth Palace Library; Library Company of Philadelphia; London Metropolitan Archives; Senate House Library; Society of Antiquaries of London Library; St. Bride Library; University College London Library; Rare Book and Manuscript Library at University of Illinois at Urbana-Champaign; University of Pittsburgh Libraries; University of Virginia Libraries; Wellcome Library; Dr. Williams's Library; and Worcester College Library.

Thanks also to my professors, colleagues, and friends who have accompanied me on this journey. Karen Parshall helped me figure out how I wanted to be a historian of mathematics, while Paul Halliday has been a font of wisdom, enthusiasm, and encouragement throughout my career. Everyone should have an advisor like Paul in their corner. I want to thank everyone who gave me encouragement, writing accountability, or feedback on my endless drafts, including the late James N. McCord Jr., Dale Hoak, the late Joseph C. Miller, Margaret Lewis, Anne Throckmorton, Jason Eldred, Heather Wiedner, Brian Dudley, Kristen McCabe, John Collins, Ryan Bibler, Sophia Rosenfeld, Michael Suarez, Erik Midelfort, Megan Brett, Abigail Mullen, Lincoln Mullen, my thoughtful peer reviewers at Oxford University Press, and my amazing editor Susan Ferber. Thanks are also due to the countless friends and colleagues who kept me company on Twitter, Slack, and other social networks along the way.

Throughout this project, generous financial support was provided by the University of Virginia Graduate School of Arts and Sciences; Folger Institute; American Philosophical Society; John "Bud" Velde Fellowship at the Rare Book and Manuscript Library at the University of Illinois at Urbana-Champaign; Dibner History of Science Program at the Huntington Library; Dumas Malone Fellowship established by the Thomas Jefferson Foundation; Carnegie Mellon University Libraries; and the Department of History and Art History at George Mason University.

Last, but never least, I want to thank my family for always being in my corner.

A Note on Style, Dating, Money, and References

Early modern punctuation, capitalization, italics, and spelling have been retained, with the exception of abbreviations, which have been expanded to their modern equivalents by the insertion of the missing letters in italics.

Unless otherwise noted, dates are given in the Julian/Old Style calendar that was used in England until 1752. With respect to the date of the new year, consistency is maintained with the original source. Although the new year officially began on Lady Day, March 25, most almanacs and printed books preferred to use January 1.

Pre-decimal English money contained 12 pence (abbreviated *d.* from *denarius*) to the shilling, and 20 shillings (abbreviated *s.* from *solidus*) to the pound (abbreviated *li.* or £ from *libra*).

When possible, references have been made to modern, printed, or electronic versions of manuscript material to facilitate ease of access. Signatures are cited when printed page numbers are mistaken or not available.

By the Numbers

Introduction

"Number, Weight and Measure": Numeracy in Early Modern England

In an almanac written for the year 1659, George Wharton mentioned a biblical passage that would have been familiar to early modern "well-wishers," "lovers," and "students" of mathematics: "For indeed all things were made by *God*, in *Number, Weight* and *Measure*."[1] An unknown reader found this passage so interesting that they underlined it in their copy.[2] The implications of this simple statement were enormous. If numbers were indeed the building blocks of God's Creation, then everything in the world could be understood through the use of numbers, particularly the concrete numerical exercises of weighing and measuring. Furthermore, a knowledge of numbers would be necessary for those who wished to study any aspect of God's inherently numerical Creation.[3] This biblical passage was not a common subject for Sunday sermons, and the Book of Wisdom in which it can be found is technically part of the Apocrypha. Nonetheless, Wisdom 11:21—God's creation of the world by number, weight, and measure—formed a rallying cry for people who wished to encourage the wider study of mathematics among the general population of early modern England.

One of the earliest and most widely published of these English mathematicians was Robert Recorde, whose *The Ground of Artes* (1543) was reprinted throughout the sixteenth and seventeenth centuries and became a byword for vernacular arithmetic textbooks. Recorde considered numbers to be the foundation, or ground, of all human arts and sciences.[4] As such, they were indispensable to every aspect of early modern society: "no man can do any thing alone, & moch lesse talke or bargayne with other, but he shall styll haue to do with numbre." Recorde expected some of his claims to be self-evident to his readers. For example, clerks, auditors, geometricians, and astronomers—"whose offices without Arithmetike is nothing"—must understand numbers as a matter of course. However, Recorde also argued that those who wished to study a wider variety of subjects, including law,

By the Numbers. Jessica Marie Otis, Oxford University Press. © Oxford University Press 2024.
DOI: 10.1093/oso/9780197608777.003.0001

grammar, and philosophy, would find themselves sorely lacking if they did not also possess some knowledge of numbers. Most important, for England as a whole, a knowledge of numbers was essential for the "gouernaunce of commyn weales in tyme of peace: and in dewe prouision and order of armes in tyme of warre." In short, "[w]herfore as without nomberynge a man can do almost nothynge, so with the helpe of it, you maye attayne to all thyng."[5]

Recorde conceived of a world that was inherently quantifiable, in which numbers and mathematics could be applied to every aspect of the world to help accomplish any conceivable goal. He wanted the mathematical arts and sciences to be not just the domain of an elite body of professionals but available to all men and women, no matter their station in life. As such, he formed part of a vanguard of mathematical educators who helped transform English numerical practices during the late sixteenth and early seventeenth centuries.

This transformation in numerical practices was complex and wide-ranging in its impact on early modern society and thought. In part, it was a transformation in symbolic systems—the culturally agreed upon symbols and syntax used to represent numbers. It was also a transformation in mathematical education, enabled by increasing literacy rates and the printing revolution. Most important, it was a transformation in technologies of knowledge, specifically the way the people of early modern England conceived of and used numbers in their daily lives.

This numerical transformation completely reconfigured the way English men and women understood the world around them. At the start of the sixteenth century, ordinary people lived in a culture where numbers were useful in certain contexts but not central to their everyday lives. Men like Recorde had to exhort people to learn arithmetic and justify why a lawyer or a naval office might care about numbers. But over the course of the next two centuries, England slowly became a place where everyone from rural farmers and urban artisans to scholarly elites and privileged nobles casually used numerical symbols and mathematical concepts to discuss abstract phenomena such as time and chance. It became a place where natural philosophers used numbers to theorize how the world worked but also where theologians used numbers to support their ideas about God and government officials used numbers to justify their policy decisions. This book demonstrates that the early modern period was a time of transition that fundamentally reoriented English ways of knowing about the world to rely increasingly on numbers.

The foundation of mathematical knowledge is numeracy. Although early modern authors examined the human ability to work with numbers, the noun "numeracy" and its associated adjective form "numerate" are only very recent additions to the English language.[6] The *Oxford English Dictionary (OED)* attributes the first use of both words to a report written by the English Ministry of Education in 1959. The words were explicitly intended to "represent the mirror image of literacy" when drawing contrasts between the reading-based humanities and the calculating-based sciences.[7] Though the concept has become more familiar in our data-driven society, and the word "numeracy" has won out against competing phrases such as "quantitative literacy," it retains its strong derivative association with literacy.

There is analytical value in contrasting numeracy—however it is defined—with literacy. Both are technologies of knowledge. Literacy has also been studied extensively by scholars over the centuries, creating possible methodological precedents. The *OED* currently defines literacy as "the quality, condition, or state of being literate; the ability to read and write."[8] While there are nuances to interpreting this definition for various times and places, in general it is understood that literacy can be judged by the demonstrated ability to read and/or write using culturally appropriate language and tools.[9] This definition provides relatively clear behavioral markers that can be used to judge whether someone has acquired the skills of literacy.

The *OED* definition of numeracy is almost—but not quite—parallel to its definition of literacy: "the quality or state of being numerate; ability with or knowledge of numbers."[10] The definition of numeracy does not reference any behavioral markers by which such knowledge can be judged.[11] In a twenty-first-century context, its vague reference to "knowledge of numbers" could be interpreted as anything from the ability to count to mastery of algebra or statistics. During the early modern period, there was a similarly wide range of standards that could be used for judging numeracy. The *OED*'s definition of "numerate" provides a bit more clarity, asserting that a numerate person is "competent in the basic principles of mathematics, esp. arithmetic; able to understand and work with numbers."[12] While this suggests that some sort of arithmetical knowledge is a key component of numeracy, it still fails to define the "basic" principles of mathematics or what it means to "understand" or "work" with numbers.

This ambiguity allows for a historical "knowledge of number" to be understood within the context of specific times and cultures. Many early modern authors were keenly interested in humans' God-given abilities with respect to numbers.[13] Mathematicians such as Recorde believed that a knowledge of numbers was one of the primary distinctions between humans and animals, arguing that those who "contemne nomber" were "as brutysshe as a beaste, and vnworthy to be counted in the felowshyp of men."[14] This distinction was so fundamental that they believed only humans could even recognize quantities in nature. According to Recorde:

> in nomber was there neuer beaste found so connyng, *that* coulde know or discerne one thynge from many, as by dayly experyence you maye well consyder, when a bytche hath many whelpes, or a henne many chyken, & lyke wayes of other, what so euer they be, take from them all theyr yonge, sauyng onely one, & you shall perceyue playnly *that* they mysse none, though they wyll resyste you in takynge them away, and wyl seke them agayne, *that* they maye knowe where they be; but els they wyll neuer mysse them truely, but take awaye that one that is lefte, and then wyll they crye and complayne: and restore to them *that* one, then are they pleased agayne.[15]

Recorde's most significant point—that the numerical abilities of humans greatly surpass that of animals—aligns with the modern science of numerical cognition, though his specific example falls apart upon close examination. We now know that elements of numerical thinking are not limited only to humans but have deep evolutionary rooms and are widespread in the animal kingdom. Many animals share the ability to subitize: to recognize immediately, without counting, the quantity of objects in a small group, usually up to four or five.[16] Various types of fish, amphibians, birds, primates, and other mammals can further discriminate between two quantities—that is, they can recognize that five is less than ten. Some animals can also perform rudimentary arithmetic and interpret ordinal information, such as choosing the third object in a list. While humans are the only species known to have the ability to create numerical symbols—that is, symbols that are assigned abstract numerical meaning—there are even a few animals such as primates and parrots that can be taught to interpret symbolic numbers.[17] Furthermore, it isn't necessary to count the number of puppies in a room to know that a specific one is gone, as indicated by Recorde's admission that dogs will search for their missing puppies. A dog investing her emotional energy into the puppy that remains to her doesn't mean that she is quantitatively unaware that the others have been taken, as Recorde asserts. While he provided only this single example, he expected his audience to have myriad "daily experiences" that would be sufficient to convince them of the truth of his ideas, whether those experiences were with farm, household, or wild animals. For Recorde, humans' numerical superiority was a self-evident truth.

John Bulwer, a medical practitioner interested in deafness and sign language, similarly argued in his *Chirologia* that "man [is] a naturall Arithmetician, and the only creature that could reckon and understand the mistique laws of numbers."[18] For Bulwer, the line between humans and other animals lay in the ability both to reckon—that is, perform addition, subtraction, or other arithmetical calculations—and to understand abstract "laws" that govern mathematics. Bulwer gave no particular examples of these laws; he could have been referencing anything from simple number properties like the commutativity of addition to the religious significance of the number 666 to complex Rules of Three. However, the idea of "laws" in conjunction with reckoning seems to indicate that Bulwer meant that humans had an introspective and abstract knowledge of numbers and arithmetic, beyond rote memorization of words and facts.

Crucially, Bulwer believed that the uniqueness of humans' ability with numbers was not an accident but rather the foundation for humanity's dominance over all the rest of God's Creation: "that divine Philosopher doth draw the line of mans understanding from this computing faculty of his soule, affirming that therefore he excells all creatures in wisdome, because he can account." The ability to work with numbers was part of the human soul and something that God had created to be the foundation of human wisdom. Consequently, humans' ability to reason actually depended on their ability to use numbers. Bulwer went on to "account such for idiots and halfe-sould men who cannot tell to the native number of their *Fingers*."[19] Those people who could not count to at least ten were intellectually deficient; they lacked the "computing faculty" part of the soul, which was necessary to be fully human.

Bulwer's use of the legal term "idiot" indicates that the lack of ability with numbers could have serious consequences in early modern England. Judge and legal writer Sir Anthony Fitzherbert defined the inability to comprehend numbers as a standard by which local officials could determine someone's mental competency to hold or inherit property:

He who shall be said to be a Sot and Idiot from his birth, is such a person, who cannot accompt or number twenty pence . . . so as it may appear, that he hath no understanding of reason what shall be for his profit, or what for his loss: But if he have such understanding, that he know and understand his letters, and to read by teaching or information of another man, then it seemeth he is not a Sot, nor a natural Idiot.[20]

Here Fitzherbert recognized a vital distinction between the ability to read and the ability to count. Literacy was a learned skill, something that could only be acquired through the assistance of some teacher. Literacy could then constitute sufficient proof of mental competency, as only an intelligent human could acquire that skill. By contrast, the ability to count was something that every reasonable human should possess. For early modern officials, the inability to count constituted sufficient proof of mental incompetency, as opposed to mere lack of education. A person who could not count to twenty could be considered an "Idiot from his birth" and naturally deficit in some mental facility common to the rest of humankind.[21]

Indeed, modern science has shown that all humans are born with a rudimentary understanding of numbers; even newborn babies can subitize and identify differences in quantities as long as there is at least a 1:3 ratio between the objects to be counted.[22] Infants also have the ability to perform approximate addition and subtraction—e.g., 25 minus 17 should result in about 5 to 10.[23] As long as they are raised in a culture that has created numerical symbols, children acquire an understanding of those symbols without much if any deliberate training.[24] Furthermore, there seems to be something about the process of mapping a mental representation of quantity onto a numerical symbol that makes the underlying abstract concept of that quantity grow more precise, enabling exact quantity comparisons and arithmetical operations as opposed to just the approximate mathematical operations enabled by evolution.[25] Thus almost all children raised in early modern England would have been able to meet Fitzherbert's numerical standard for mental competency.

Recorde, Bulwer, and Fitzherbert made several varying assumptions about humans' natural knowledge of numbers. All three took for granted that there were spoken number words with which every English speaker should be familiar—one, two, three, and so forth. Indeed, the most common numerical symbols employed across human cultures are linguistic, though in some cultures gestures appear to serve the same function as number words.[26] Despite the intermediary role played by language in most forms of number words, MRI experiments have shown that number words and other mathematical exercises activate areas of the brain that evolved to deal with quantity rather than non-mathematical language.[27] Crucially, despite the common intermediary role played by words, the human understanding of quantity is abstract and largely independent of language.

These three authors also took for granted that spoken number words were distinct from all other words in the English language. This is not true of every early modern language. For example, Bulwer's sign language in the *Chirologia* did not have separate number gestures; instead it created a one-to-one correspondence between numbers and the letters of the alphabet.[28] By contrast, someone learning English had to learn the number words separately and could not repurpose an existing word for counting.[29] Learning a number word in English required the learner not only to memorize the word and its pronunciation but also to demonstrate their understanding of the existence of an abstract quantity indicated by that word.[30] Modern children go through these steps sequentially; they usually acquire the basics of their culture's linguistic symbols for numbers alongside other words in their first language(s) and subsequently learn to associate those symbols with the abstract quantities the symbols are meant to represent.[31] This doesn't happen all at once and children across cultures go through consistent intermediary steps: first, they understand the quantity 1; then they understand 1 and 2; then they understand 1, 2, and 3; then they understand 1, 2, 3, and 4, and make the cognitive leap to understanding that each successive number word indicates one more item than the previous one. From this basic understanding of plus or minus 1, most children then figure out how to add and subtract without further instruction. Even at this stage, their understanding of quantities is abstract: bilingual children who acquire the understanding that "three" means 3 almost immediately make the connection that "tres" also means 3.[32]

One of the largest cultural differences at this point in a child's acquisition of number symbols derives from the numerical base used in the culture's number system. The counting numbers in early modern England—which are also the dominant counting numbers used around the world today—are decimal or base 10 numbers, in which all numbers are formed by a combination of a power of 10 and a number less than 10.[33] Regardless of the base number, the culture's language will develop a series of lexical number words (e.g., in the English-language base 10 system these are "one" through "nine") as well as words for powers of the base (e.g., ten, hundred, thousand, million) which can be combined syntactically to generate an arbitrarily large—though not infinite—set of counting numbers.[34] The extent to which a culture's number words correspond to their chosen base influences the speed at which children learn place-based values, and thus the ability to syntactically form larger number words. English number words map relatively well to a base 10 system, but there are several irregularly formed number words, particularly between 10 and 100.[35] For example, in a perfectly regular mapping, the number 11 ought to correspond to a syntactical word meaning ten-one (10 + 1) but in English it is the word "eleven."[36] As with modern English-speaking children, these slight variations would have delayed early modern English children's acquisition of numerical syntax.

This is important because where Recorde, Bulwer, and Fitzherbert differed was in their assumptions about the extent of man's natural ability to string together these number words in a given sequence. Recorde only explicitly referenced the ability to differentiate one and more-than-one, though he strongly implied the ability to count an unspecified number of objects, presumably up to the limit inherent in then-extant English number words. This demonstrated an understanding of the associated quantities by creating a one-to-one relationship between number words and physical objects. Bulwer similarly required an accused idiot to count, but only to ten, the number of most humans' fingers, arguing that "all men use to count forwards till they come to that number of their *Fingers* . . . [which] by an ordinance of nature, and the unrepealable statute of the great Arithmetician, were appointed to serve for casting counters" for humanity.[37] Bulwer's formulation leaves unclear whether one needed to understand ordinality—to be, in modern scientific jargon, a cardinality-principal-knower who understands that counting to 10 results in a collection of 10 items—or merely recite number words with the mnemonic aid of fingers.

By contrast, Fitzherbert's legal text picked twenty as the smallest number to which his alleged idiot must be able to count. "One" through "twenty" form the complete set of lexical and irregularly formed English number words in the counting numbers.[38] Beginning with "twenty-one," the rest of the English number words are syntactical, formed by combinations of powers of ten and numbers less than ten. The syntax grows more complicated with the addition of hundreds, thousands, tens of thousands, and so on, but the numbers remain syntactical in nature.[39] Thus Fitzherbert required his accused idiot to demonstrate a facility with all the non-syntactical numbers but left unspecified whether mastering the syntax of larger numbers requires an equivalent intellect to mastering the symbols and syntax of written language.

Beyond counting, Recorde, Bulwer, and Fitzherbert also assumed that all mentally competent human beings would be able to perform varying degrees of arithmetic, particularly addition and subtraction. Recorde began the arithmetic section of his textbook with the addition of three-digit numbers and assumed his readers would be able to mentally add and subtract one- or two-digit numbers such as 6 plus 8 or 14 minus 12; he also refused to teach multiplication of "small dygetes vnder 5 . . . seynge they are so easy that euery chyld can do it."[40] Bulwer's comments about human arithmetical ability lacked specific details, but Fitzherbert clearly tied human arithmetical skills to accounting, requiring the ability to differentiate profits from losses. In practice, mental competency examiners seem to have begun by asking people to count with small numbers such as 7, 12, or 20, then required those who succeeded in these initial tasks to work with larger numbers and/or perform arithmetic in financial contexts. For example, during a 1628 examination, a man was asked to subtract £120 from £200, while other accused idiots were required to compare or convert quantities across currencies, which required either serial addition and subtraction, or multiplication and division.[41] It is unclear whether this arithmetic had to be performed mentally or whether it could be performed using verbal counting strategies or some physical aid. However, it appears to have been the abstract ability to add and subtract that was important, not the means by which this task was accomplished.

Modern science has upheld many of the assumptions these authors made about human numerical abilities at the aggregate level, within the specific context of English-speaking cultures. However, it is vital to remember that individuals within the same culture vary in their aptitude for numbers. This is particularly important given the scarce evidence available for certain historical cultures and the associated need to infer generalized information from a small number of examples. Modern scientists have calculated that somewhere between 3% and 7% of humans have dyscalculia—a severe difficulty learning basic numeration and arithmetic—which is linked to deficits in their inborn abilities to assess and estimate quantities.[42] Dyscalculia can manifest as poor fact retrieval, failure to master spatial concepts, and/or weak understanding of arithmetical procedures.[43] Variations in people's inborn ability are also predictively related to the types of strategies they adopt to perform mental arithmetic, including addition, subtraction, and multiplication.[44] When one is reconstructing historical numeracies, individual variations should not distort conclusions about broader trends.

In early modern England, the ability to learn English number words, along with some level of arithmetical skill, was considered inherent in every mentally competent human being and something every child should learn without much, if any, deliberate instruction. If this is used as the marker for an early modern "knowledge of number," then the entire population of England was almost universally numerate and had been for centuries. While such a result might be emotionally satisfying, it obscures very real differences in numerical practices within the population and across the early modern period. It is therefore more useful to consider the knowledge of English number words as the baseline for a range of numerical skills whose acquisition might indicate a "knowledge of number."

As Bulwer's references to finger-counting and casting counters implied, number words were only one way of symbolizing numbers in early modern England. Other symbolic systems had to be learned separately from the English language, through formal or informal training, and varied widely. Some were performative, such as hand gestures. They required no physical aids but were also incapable of being used to communicate over large physical or chronological distances. Others were based on the manipulation of objects, like tally sticks and counters. The materiality of these objects created the possibility of an object also being imbued with both numerical and non-numerical information. Still others were written, like Roman and Arabic numerals. Mastery of these systems required access to a specific set of tools as well as prerequisite skills associated with literacy. While written systems are most familiar to and privileged by modern scholars, there was no dominant system during the early modern era. Several of these systems were, in practice, codependent with other systems.

To accommodate this variety of symbols, early modern English numeracy can be provisionally defined as knowledge of a symbolic system for representing and manipulating numbers that is distinct from the English number words. While historically sensitive, this definition lumps together a wide variety of symbolic systems and obscures the differences between them. It is therefore useful to additionally treat each symbolic system as its own type of numeracy—e.g., hand gesture numeracy, tally stick numeracy, etc.—and to recognize the existence of early modern English numeracies, plural, within the overarching formula of early modern English numeracy, singular. This multiplicity allows for analysis of the differences between various "knowledges of number" and the ways in which these differences shaped how people chose to use each symbolic system. Such choices were not neutral decisions. Indeed, they profoundly influenced what and how information could be conveyed in early modern England.

———

By the Numbers explores the quantitative transformation of early modern England, particularly its relationship with numeracy and religion. It does not attempt to quantify historical numeracy rates but rather focuses on describing early modern numerical practices and analyzing how numbers came to matter so much—and be so widely embedded—in English culture. The first part of this book establishes the variety of early modern English number symbols, including their material characteristics and how ordinary people learned to use them, while the second part consists of three case studies that place numbers in conversation with broader developments in English history.

Chapter 1 begins by examining the universality, multiplicity, and materiality of early modern numeracies. Prior to the modern dominance of Arabic numerals, the people of sixteenth- and seventeenth-century England utilized a variety of different symbolic systems to record and manipulate quantitative data. Common systems included performative, object-based, and literate options, and people's choice to use any particular system was informed by that system's material characteristics. While early modern patterns in Arabic numeral adoption have confused many scholars, they are in fact consistent with a culture where multiple forms of numeracy coexisted without a single dominant system. Over the course of the early modern period, English men and women adopted Arabic numerals in a context-specific fashion, commonly employing them side by side and interchangeably with other symbolic systems.

During the late sixteenth and seventeenth centuries, many English account-keepers began to replace their existing object-based and literate arithmetical practices with new practices based on Arabic numerals. These men and women considered the material characteristics of symbolic systems when making decisions about whether or not to trust a system to perform either of the conceptually distinct functions of recording and calculation. In particular, they looked at the permanence of a system's symbols when choosing a system for recording, and they developed strategies to increase symbols' resistance to after-the-fact alterations. At the same time, they looked for the opposing characteristic of manipulability when choosing a system that would facilitate calculations. Because recording and calculation remained conceptually distinct, people could employ different systems for each function, choosing the optimal system for the task at hand. Although Arabic numerals had the potential to perform both functions, they were not the optimal solution for either one. As a consequence, Arabic numerals were neither immediately nor universally believed to be a significant improvement over other early modern symbolic systems, despite being the only extant system with the potential to unite recording and calculation. Chapter 2 examines both the incentives and disincentives to Arabic numeral adoption, critiquing the common modern tendency to assume their inherent superiority.

Just as there were multiple symbolic systems in early modern England, there were multiple ways to acquire a knowledge of them. England's increasing literacy rates and the development of vernacular arithmetic textbooks were essential to changing arithmetical practices in this period. Textbooks were better suited to teaching written, rather than object-based, methods of arithmetic and enabled individual educators to reach wider audiences than face-to-face methods of instruction. By exploring the qualities of printed books, analyzing marginalia in arithmetic textbooks, and examining changing educational advertisements and curricula over time, Chapter 3 demonstrates the importance of literacy and didactic literature to early modern arithmetical education.

Chapter 4 demonstrates how the methods early modern people used to locate themselves in time became more complex and abstract over the course of the early modern period, even as time retained its religious significance. Politically and religiously contested calendar reforms led English men and women to adopt the mathematical notion of fractions, while the proliferation of almanacs and increasing global travel raised popular awareness of both the relativity of time and its interconnectivity with space. By the eighteenth century, English men and women understood calendars and times to be primarily numerical constructs; calendars were not sacred in and of themselves, but rather tools people could employ to conduct both their mortal and immortal lives.

The second case study explores early modern conceptions of quantified chance and its relationship to God's providence to show how early modern people used numbers in an attempt to cope with the uncertainties of the future. Over the course of the sixteenth century, gamblers adapted the existing language of proportional odds to refer to a quantified version of chance, which the wider population subsequently began to apply to situations in their everyday lives. This brought quantified chance into potential conflict with established views of divine providence by the turn of the seventeenth century. Chapter 5 demonstrates how people increasingly embraced the concept of using numbers to predict their futures and minimize risks, even as they maintained their belief in an all-knowing, all-powerful God. It was God who made the world to function by natural, numerical laws, and thus it was because of God's providence that people could use those numbers to calculate the likelihood of future events and avoid the undesirable ones.

The final case study in Chapter 6 examines how new ways of thinking with numbers led to the transformation of early modern censuses from the creation of lists in support of government fiscal-military initiatives to the aggregation of the necessary data for quantitatively analyzing populations in support of governmental policymaking more generally. During the sixteenth century, census data were most commonly collected in qualitative list formats and applied toward immediate policy goals. However, over the course of the seventeenth century, people increasingly used the collection and quantitative analysis of demographic data as evidence to support social and political arguments. This led to the post-Restoration articulation of "political arithmetic," in which numbers were manipulated to support monarchical policy goals. Later practitioners reimagined their work as the analysis of demographic data whose power to influence politics lay in its avowed neutrality, transforming political arithmetic from a method of governing by numbers to a more general method of reasoning by numbers, which could be applied to a host of political, economic, and social questions.

As these diverse examples demonstrate, reconstructing the cognitive element of numeracy is vital for understanding the wider culture of early modern England. Only by examining numbers through the lens of early modern English culture—and, crucially, defamiliarizing the Arabic-numeral numeracy we so often take for granted—can we comprehend the full range of early modern numeracies. Recovering these numeracies illuminates a revolutionary shift in the way people used numbers during the early modern period. Changes in symbolic systems and educational practices encouraged new ways of thinking, as people began to employ numbers as a tool for interpreting the world around them. The men and women of sixteenth- and seventeenth-century England never doubted that God created their world, and at the turn of the eighteenth century, they still lived in a world made by God. But it was also a world made—and made understandable—by number, weight, and measure.

1

"The Dyuers Wittes of Man": The Multiplicity and Materiality of Numbers

In 1607, an entry in the Earl of Northumberland's accounts recorded 20s. paid out "to one that taught the accountant, Mr. Fotherley, the art of arithmetic."[1] At first glance, this appears nonsensical. The accountant in question was already keeping the earl's books and must have known how to signify quantities, add, subtract, and more. The multiplicity of early modern numeracies stands at the root of this terminological confusion. Mr. Fotherley was not learning the basics of arithmetic for the first time but rather how to perform arithmetic with a new symbolic system of numbers. Given that this occurred at the turn of the seventeenth century, the tutor was most likely instructing him in the still relatively new written method of arithmetic with Arabic numerals.

Early modern English men and women utilized a wide range of symbolic systems to record and manipulate quantitative data, including but by no means limited to the written systems preserved in account books. In his 1543 arithmetic textbook, Robert Recorde covered three of these systems, then informed his readers that he couldn't possibly teach them every system they might encounter. They must learn other systems by observing "the workying of eche sorte: for the dyuers wittes of men haue inuented dyuers and sundry wayes almost vnnumerable."[2] Given this variety of systems, Recorde saw no reason why those with little other learning—particularly "them that can not write and rede"—shouldn't be able to acquire knowledge of one or more symbolic systems and use arithmetic in their day-to-day lives.[3]

The main symbolic systems employed during the sixteenth and seventeenth centuries can be roughly divided into three categories based on their most prominent material characteristics: performative, object-based, and written. Performative systems did not require the assistance of external physical objects and could be employed under any circumstances. Object-based systems involved the alteration or manipulation of physical objects, which were invested with numerical meaning that persisted across time and space.

By the Numbers. Jessica Marie Otis, Oxford University Press. © Oxford University Press 2024.
DOI: 10.1093/oso/9780197608777.003.0002

Written systems also involved the assistance of physical objects—commonly pen, ink, and paper. However, these objects were only a means to create and convey numerical symbols, not symbols themselves; the same symbols could be drawn on wax tablets or dust boards, engraved into wood and stone, or embroidered onto cloth. Although these categories are analytically useful, it is important to recognize that most systems had multiple material characteristics; for example, tally sticks could be both notched to record quantities as well as written on, to indicate the names of parties in a financial transaction. The boundaries between these categories should therefore be considered fluid rather than hard-and-fast divisions.

Humans naturally transcode—that is, translate back and forth between the abstract idea of a specific number and numerical symbols—when they acquire number words. Learning additional symbolic systems further develops that ability by associating even more symbols with each abstract concept. Numerical information can further be transcoded between symbolic systems, such as a twenty-first-century American mentally translating the dot arrays on dominoes into Arabic numerals to be written on a game's scorepad.[4] However, numerical symbols are not merely neutral conveyances for quantitative information; they are layered with additional capabilities and meanings by their material characteristics and the cultures that invent and use them. For the men and women of early modern England, their choice to use any particular numerical system was informed by both their cultural norms and each system's material characteristics.

Words and Gestures: Embodied and Performative Symbolic Systems

The most universal symbolic systems in early modern England were spoken words and gestures, which enabled people to count and calculate without any physical aids other than their minds and bodies. Oral numeration, in particular, was something that all mentally competent human beings were supposed to learn along with language. Robert Recorde began his arithmetic textbook, *The Ground of Artes*, with a discussion between a master and scholar that mocked the concept of speech without numbers. When the scholar complained about having to learn arithmetic, the master challenged him: "Yf nombre were so vyle a thynge, as thou dyddest esteme it, then nede it not to be vsed so moch in mens communycation.

Exclude nombre and answere me to this question: Howe many yeares olde arte thou?" The scholar was only able to respond with "Mum," an inarticulate sound indicating refusal or inability to speak. The master continued to harry the scholar with unanswerable questions: "How many dayes in a weke? How many wekes in a yere? What landes hath thy father? How many men doth he kepe? How longe is it syth you came from hym to me?" He mocked the scholar's continued inability to provide a numberless answer: "So that yf nobmre wante, you answere all by mummes." Finally the scholar stopped playing by the rules and responded to a question about the distance to London with the nonsensical statement: "A poke full of plumbes." The master's point was made.[5]

Oral numeration formed the basis of early modern English numeracy. The numerical information from other symbolic systems could always be transcoded to and communicated through spoken number words. Philosopher Thomas Hobbes even believed that non-verbal symbolic systems were useless without this verbal reference base. According to him, words registered human thoughts and "without words, there is no possibility of reckoning of Numbers."[6] Educators such as grammar school master John Brinsley encouraged the connection between spoken number words and other symbolic systems by requiring students to practice transcoding between them. Brinsley had his scholars read and recite numbers "backwards and forwards, so that your scholler be able to know each of them, to call them, or name them right, and to finde them out, as the child should finde any letter which he is to learne: in a word to tell what any of these numbers stand for, or how to set downe any of them."[7]

After children learned number words, arithmetical calculations could be performed mentally, verbally, or using fingers for physical assistance. John Hall of Richmond described children learning to add small numbers as a combination of all three processes. "We finde children, one with another, making it one of the first tryals of their abilities to pose each other in mental addition of numbers," by first becoming 1-, 2-, 3-, and 4-knowers then "having the question asked them how many two and three do make, or the like" small numbers. The child so questioned would respond by "calling into memory the figure of any three and two things so and so posited; and so, by comparing them do know what they amount unto." When children were unable to rise to the challenge, they resorted to a physical aid to discover the answer: "they help themselves in their numeration by an outward figure, as by counting on their fingers, or the like."[8]

Hall's observations of seventeenth-century English children's early arith-metic strategies parallels observations of twenty-first-century American children's strategies. By kindergarten, most American children can solve simple addition and subtraction problems, and they employ strategies such as counting verbally or on their fingers to support their arithmetic. One such strategy for adding two numbers, e.g., 3 and 5, involves counting 1, 2, 3 then "counting on" 5 more numbers (4, 5, 6, 7, 8) to reach the answer. A faster strategy involves starting at 3 and just counting 5 more numbers, or realizing that the process will be fastest if they start with the largest number and so beginning at 5 and counting 6, 7, 8 to the answer. Such strategies lead to children eventually being able to directly retrieve the answer to an arithmetical question, such as simply "knowing" that 3 + 5 = 8, and to use decomposition strategies—e.g., knowing 6 + 6 = 12 so 6 + 7 must be 13—to extend their arithmetical knowledge.[9]

When mental manipulation of numbers becomes too difficult, humans naturally transition to using body parts as an aid to counting and calculation. Modern children appear to pick up this behavior spontaneously by modeling behavior they see in older children or adults.[10] In early modern England and Europe, the body parts in question tended to be limited to fingers and usually involved putting quantities in one-to-one correspondence with fingers; indeed, the English word digit derives from the Latin word for finger.[11] However, other systems are possible. The eighth-century English monk Bede developed a specialized system for calculating solar and lunar cycles that involved counting on a combination of fingers and finger joints, and a similar system was used by navigators in the early modern period to calculate the age of the moon and the current state of the tides.[12] More recently, the Torres Strait Islanders use fingers, arms, torso, legs and toes in a specific order to denote numbers up to 33, while the Oksapmin-speaking people of New Guinea count to 27 using their fingers, shoulders, and eyes.[13]

Both modern scholars and early modern thinkers such as Hobbes attribute the origin of base 5 and 10 arithmetical systems to finger counting.[14] As Hobbes explained:

> there was a time when those names of number were not in use; and men were fayn to apply their fingers of one or both hands, to those things they desired to keep account of; and that thence it proceeded, that now our numerall words are but ten, in any Nation, and in some but five, and then they begin again.[15]

The overwhelming majority of historical symbolic systems are base 10, and finger counting is the likely ultimate origin of this similarity. However, language is an important intermediary factor; in all known cases of independently invented symbolic systems for numbers, the system has the same base as the number words in the culture's native language.[16]

While it isn't possible to quantify the extent to which people finger counted in early modern England, anecdotal evidence suggests it was a widespread practice among both children and adults. Physician John Bulwer believed finger counting to be a universal experience "by all Nations," as fingers were "quicke and native digits, alwaies ready at *Hand* to assist us in our computations." The method Bulwer described mirrored the left-to-right reading direction of English and other European languages:

> to begin with the first finger of the left hand, and to tell on to the last finger of the right, is the naturall and simple way of *numbring* & *computation*: for, all men use to count forwards till they come to that number of their *Fingers*.[17]

For Bulwer, the perceived universality of this method was no accident, but part of God's express design, for humans' ten fingers were "those numbers that were borne with us and cast up in our *Hand* from our mothers wombe, by Him who made all things in number, weight & measure."[18]

Unlike in later centuries, there appears to have been no stigma attached to the adult use of finger counting during the early modern period.[19] A late seventeenth-century edition of Diego Hurtado de Mendoza's *Lazarillo* had the title character "counting on my Fingers, how I should lay out my Money" as he ran to market, while earlier in the century, Pierre de La Primaudaye argued that it was short-sighted and foolish for men to "marrie by the report of their fingers, counting vpon them how much their wiues bring to them by mariage."[20] Although these were likely references to straightforward addition and subtraction, some mathematicians taught students to perform even the relatively complicated operations of multiplication and division on their fingers. At the turn of the eighteenth century, Scottish mathematician and minister George Brown considered "how to add, subtract, multiply, and divid by the figners [sic]" to be preliminary knowledge for those wishing to learn decimal arithmetic. He gave no explanation for adding numbers less than six, which he considered intuitive, and his method for adding numbers greater than five was almost as simple. Given two numbers, "you may make the one on the left, the other on the right Hand, accompting the shut hand five; and if you open one Finger 6, two 7, three 8, and four 9."[21] The sum was ten plus the sum of the open fingers. His methods for multiplication and division of numbers over five were more complicated, often requiring the calculator to carry a one from the units place to the tens place; this was not a system for someone unfamiliar with the basic principles of arithmetic.

Finger counting sometimes even seems to have been seen as a sign of or proxy for arithmetical skill during the seventeenth century. One of author Margaret Cavendish's characters enters a library and encounters several men, each carrying an object or gesturing in a manner that indicated their subject of study: law, arithmetic, astronomy, geography, moral philosophy, theology, natural philosophy, and poetry. One man was "counting on his Fingers, and looking in his Book; by which he saw he was studying Arithmetick."[22] Conversely, philosopher Mary Astell indicated to the readers of one of her treatises that they should mock the ignorance of her "city-critick" character by describing him as someone who aspired to arithmetical learning but "mistakes frequently in the tale of his Fingers."[23] Beyond the literary world, John Bulwer described statues of Arithmetic embodied in human form— including one "in the new Ovall Theater, lately erected for the dissecting Anatomies in Barber-Surgeons Hall in London"—that were visually designated as Arithmetic by hands sculpted to look as if they were engaged in "Manuall Arithmeticke."[24]

While it is likely that most people in early modern England were simply placing numbers in one-to-one correspondence with their fingers, Bulwer's statues were using a more formal method for finger counting that was in widespread use throughout the Mediterranean during the Roman Republic and early medieval periods. There are currently no known descriptions of how the calculations worked, but visual depictions of this method of finger counting appear on ancient Greek and Roman reliefs, mosaics, and tesserae. The earliest surviving treatise that describes the finger counting method in full—up to its presumed maximum of one million—is the 688 *Romano computatio*. The method also appears in Bede's *De tempore ratione liber* in 725. This English monk's treatise on reckoning time helped popularize finger counting as a silent method of communication among the monks of northern Europe.[25] It's unclear how common this method remained in the late medieval period, but it was well-known enough among Italian mathematical elites for Leonardo of Pisa to include it in his 1202 *Liber Abaci* as a physical aid for the mental calculations required in Arabic numeral arithmetic, and for Luca Pacioli to include it in his 1494 *Summa de Arithmetica*. The method also survived in the Byzantine East and can be found in two fourteenth-century letters on arithmetic by Nicolas Rhabdas of Smyrna.[26]

By the early modern period, however, the Greek and Roman finger counting system appears to have fallen out of general use in England. Robert Recorde reproduced it in his 1543 *The Ground of Artes* and extolled "the arte of nombryne on the hand, with diuers gestures of the fyngers, expressynge any summe conceaued in the mynde" but admitted that "[t]his feate hath ben vsed aboue 2000 yeares at the leaste, and yet was it neuer comonly knowen, especyally in Englysshe it was neuer taughte yet." Still, Recorde believed this gestural system was "ryghte worthy to be well marked" and suggested its "straungenes and secretnes" could make it useful for facilitating secret and silent communication.[27] His book's later editors were not impressed by a system that was only useful as a curiosity or a code. While the system was reprinted in subsequent editions through the 1590s, it was dropped from all seventeenth-century editions to make room for other, more widely marketable content.[28]

Tally Sticks and Counters: Object-Based Symbolic Systems

Although oral and gestural systems were the most universal symbolic systems in early modern England, systems based upon the manipulation of physical objects were almost as widespread. It was only a short conceptual step from fingers to external objects that, like fingers, were a manipulatable substitute for the objects to be counted. Many of Recorde's "dyuers and sundry wayes" of enumeration and arithmetic were probably object-based, as virtually any collection of objects could serve as an aid to counting and arithmetic, particularly objects small enough to carry. Exciseman John Cannon had a grandfather who used beans to keep his accounts, while Daniel Defoe knew a country shopkeeper who performed arithmetic with spoons, and the servants of George Purefoy used twigs to keep track of how many people he helped with his charity.[29] While in theory any object could be used to create numerical symbols, in practice there were two commonly known and widely used object-based systems during the early modern period: tally sticks and counters.

Humans began keeping tallies by notching bones somewhere between 35,000 and 10,000 BC, and cross-cultural studies have demonstrated the worldwide use of tallies in a variety of mediums to the present day.[30]

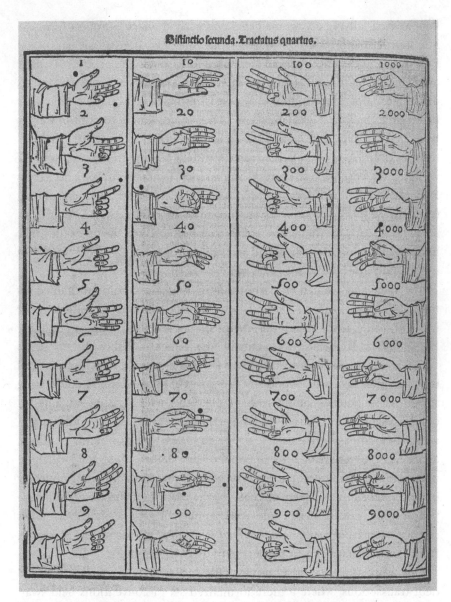

Figure 1.1 Gestural Enumeration from Luca Pacioli's 1494 *Summa*. Bodleian
Library Savile P 8. Used by permission of the Bodleian Libraries, University
of Oxford, under a Creative Commons Attribution-NonCommercial 4.0
International License. https://digital.bodleian.ox.ac.uk.

People used tallying across medieval and early modern Europe—from the Mediterranean to Scandinavia and from Russia to the British Isles—and European travelers noticed and compared foreign tallying practices to their own.[31] An English translation of Inca Garcilasso de la Vega's chronicles described and admired quipu, while John Smith noted the native Americans in Virginia "by little sticks will keepe as iust an account of their promises, as by a tally."[32] Travelers to Asia made similar observations. As early as the thirteenth century, Marco Polo compared Chinese tallying to Italian practices, while in the seventeenth century Englishmen Samuel Purchas and Lewes Roberts wrote about merchants in Vociam (modern Baoshan) and Cambalu (modern Beijing) who used a system of making "contracts and obligations in *tallies* of wood, the halfe whereof the one keepeth, and the other the other halfe, which being afterwards paid and satisfied, the said *tallie* is restored; not much unlike the custome of *tallies* in *England*."[33]

The men and women of early modern England commonly tallied by marking chalk lines on slate or by cutting notches into sticks; this latter method dated back to at least the Anglo-Saxons, and the original name of the English Royal Exchequer was actually "the Tallies."[34] Although tally sticks varied widely in size and even shape, each was a wooden stick—often hazel or some other easily split wood—with a series of notched marks cut into the sides.[35] These notches could form a simple count, with each notch in one-to-one correspondence with a thing being counted, or the notches could vary in size to function as symbols for specific quantities, such as pounds, shillings, and pence. Tallies created for personal or private use could vary widely in notational conventions—particularly when used to count non-monetary objects such as trees, sheep, or coal—to the point that sometimes only the people who created them could read them.[36] In general, however, tallies seemed to have been sufficient to meet the needs of a wide variety of people, from shepherds to alehouse keepers to tradesmen to miners.[37]

One especially widespread use for tally sticks was in tracking credit and debt transactions, where a more formalized set of conventions held sway. Two parties would cut notches into a stick then split it lengthwise, with each party keeping one half to prevent the possibility of fraud. The creditor's half was called the stock, which eventually gave its name to financial concepts including government stock, stock jobbing, and the stock exchange.[38] The debtor kept the slightly shorter half of the stick, called the foil. When the two parties reconvened to settle the debt, the two halves of the tally stick were compared and only if they matched each other—proving that neither party had attempted to defraud the other—was the debt considered paid. The tally stick could then be broken, burnt, or otherwise disposed of at will.

These split tally sticks were used by the Exchequer to keep track of its financial transactions from at least the eleventh century through the beginning of the nineteenth century. Exchequer tally sticks were so ubiquitous that many people also adopted the Exchequer cutting and labeling customs—which remained essentially fixed for over seven hundred years—in the tally sticks they used for private lending and debts.[39] These customs included a distinct notch type and location for each monetary unit, or large multiple of a monetary unit, resulting in a versatile object that could function simultaneously in all three numerical bases of the English monetary system: base 10, base 12, and base 20.[40]

Object-based symbolic systems, like tally sticks, have often been viewed as the province of the illiterate. As early as 1600, one of William Shakespeare's historical plays contrasted his character's ability to write his name with the "score and the tally" used by his "ancient forefathers," while by 1714 Roger North declared tallies "well known, being of ordinary Use in keeping Accompts with illiterate People, and serves well enough for meer Tale" but were "obsolete" for the gentleman accountant.[41] Private tallies probably catered more to the illiterate, but they do not survive in large enough numbers for firm conclusions to be drawn about their use. By contrast, the Exchequer tally sticks contained a written component that included the names of the parties originally involved in the transaction, a feature that assumes the literacy of its user base. This writing was originally in Latin—or, in a high medieval branch of the Exchequer dedicated to Jewish cases, in Hebrew—which required users to have not just literacy

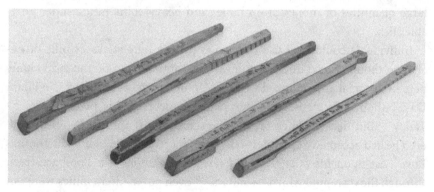

Figure 1.2 Fifteenth-century Exchequer tally sticks © Science Museum Group.

but also some ability to interpret a second language.[42] The Exchequer only issued tally sticks with English writing under the Commonwealth; in 1654 Lord Protector Oliver Cromwell ordered tallies to be "levyed and allowed according to the usual and accustomed Course of the Exchequer, with the Alteration of the Words upon the Tally from Latine to English" to make them more widely legible.[43]

The writing on Exchequer tally sticks was a vital component of the tally as a physical object, supporting its use in securing credit and debt transactions while also enabling it to effectively function as a bill of exchange. As early as the 1290s, instead of keeping large amounts of specie on hand, the Exchequer would cut a tally stick and then give the foil to the debtor and the stock to their own creditor.[44] For example, one fifteenth-century tally stick was entered into the Exchequer Receipt Roll as having been received on July 11 in "Sussex. From John Perpount and John Yerman, collectors of the king's customs and subsidies in the port of the town of Chichester, £vij of the said custom and subsidies." A marginal note was later added, "for Lord Bourchier by return of one tally, levied on the xvij⁰ day of February in the xxiij⁰ year of the present king, by the hand of Richard Wode."[45] This tally stick recorded a £7 debt owed by the king's officers, which was reassigned to cover a debt to Henry Bourchier, who likely took the stock to Chichester as evidence that they should pay him directly. Creditors did complain about the inconvenience of this system, but from the Exchequer's point of view it was ideal.[46] By this means, the Exchequer could avoid both collecting its own debts—that expense and hassle being passed off to its creditors—and disbursing specie in large quantities. This ability to cancel debts owed with the debts of others was crucial to the smooth functioning of the economy in an era with chronic shortages of available coins.[47] As an added benefit, it could also transform large quantities of money from bulky and heavy coins into a more portable stick.

Individuals could also use their own private tally sticks as bills of exchange, though such tallies' usefulness was contingent upon the individual's perceived social and financial worth.[48] Upon taking office in 1595, William Day, bishop of Winchester, easily paid the debts of his predecessor "by Talley."[49] But decades later, civil servant and diarist Samuel Pepys discovered he had accepted a worthless tally from merchant and politician Thomas Povey, as "upon his tally [I] could not get any money in Lombardstreet, through the disrepute which he suffers."[50] Even government tallies were not immune to losing their value when the perceived reliability of their issuer decreased, as goldsmiths and other tally-holders found to their dismay when King Charles II temporarily stopped payments from the Exchequer in 1672.

A logical extension of using tallies as bills of credit—and the proximate cause of the so-called Great Stop of the Exchequer—was the late medieval and early modern practice of allowing the Exchequer to cut tallies of recompense in anticipation of future revenues.[51] By the Restoration, the Exchequer was spending income well in advance of its collection by persuading creditors to accept discounted or interest-bearing tallies for debts that would not come due for weeks, months, or even years. But the cost of such advanced funds was high. Although Restoration interest rates were legally fixed at no more than 6%, Pepys often complained in his diary about the money Charles II was losing by paying fees and 10% interest rates, which was still better than 20% and 30% interest Charles would end up paying a few years later, or the 30%, 40%, or 60% interest the poor were allegedly charged by tally men, in the early modern equivalent of credit card debt and payday loans.[52] The situation only worsened after the 1672 stop of Exchequer payments. By the 1680s, Parliament attempted to curtail the king's use of tallies by resolving that anyone who paid for the creation of or accepted a tally of anticipation would be "be adjudged to hinder the Sitting of Parliaments, and shall be responsible therefore in Parliament."[53] There was even an established market for the buying and selling of government debt through tallies and other bills, a proto-stock market run by "stock jobbers" and "tally jobbers." Only the 1694 founding of the Bank of England—which was offered favorable terms to buy up government tally sticks that traded with average discounts of 40% off their face value—reined in the tally-jobbers and stabilized tallies' worth.[54] The bank accepted deposits of tally sticks worth over £112,000 in its first week alone, and the scarcity of specie encouraged the continued, extensive use of tally sticks in their first decades of existence.[55]

Tally sticks remained useful at the beginning of the eighteenth century, but their usage declined precipitously in favor of paper bills and notes over the course of the eighteenth century. In 1783, the Exchequer officially adopted indented cheque receipts—whose cut edges mimicked the split in tally sticks—and began to phase tallies out of service.[56] Tally sticks were technically abolished in the same act that introduced the cheque receipts, though the implementation of the act was officially delayed until the two incumbent chamberlains left office in 1826. Subsequent attempts to dispose of the now-obsolete tally sticks led to the Houses of Parliament burning down on October 16–17, 1834, making them ex post facto weapons of math destruction.[57]

Besides tally sticks, the other widespread object-based symbolic system in early modern England relied on sets of metal discs called counters, compters, or jettons, which were often used in conjunction with a counting board, table, or cloth to form an abacus.[58] This system had antecedents in the ancient Greek and Roman pebble abacus, which employed small stones—as well as discs of bone, baked clay, horn, and wood—for counters.[59] Various forms of the English word calculate derived from "calculus," the Latin word for both a pebble and a counter, and its associated verb "calculare," which means "to calculate" and can be transcribed literally as "to pebble." The term abacus itself came from a Latin derivation of the Greek word "abax," meaning "a flat surface" on which to perform those calculations.[60] By no later than 1110, the abacus system—specifically a "chequered" cloth used as a surface for calculations—had given its name to the English Royal Exchequer, though the adoption of metal counters was a slightly later development.[61]

The oldest minted, metal counters that can be definitively dated belonged to Blanche of Castile, queen of France at the beginning of the thirteenth century, though it is possible metal counters were also in use by northern Italian bankers and merchants at this time. Over the next hundred years, people in France, England, and the Low Countries increasingly adopted the use of metal counters, even as Italians began to abandon the abacus in favor of Arabic numeral arithmetic. During the fifteenth century, metal counters were commonly used from Scandinavia to the Iberian peninsula and from Scotland to Poland, but Venetian scholar Ermolao Barbaro felt comfortable denigrating the abacus as a system only used by foreign (i.e., non-Italian) barbarians.[62] Most fifteenth- and early sixteenth-century counters were minted in the city of Tournai—in northern France until its conquests by first Henry VIII then Charles V—which had a royal monopoly on their production, but the Free Imperial City of Nuremberg became the new center of early modern counter production in the sixteenth century. For the next two hundred years, Nuremberg continued to export counters to England and the rest of western Europe, and from there to European colonies in the Americas.[63]

Early modern counters were generally round, coinlike objects that varied in size and metallic composition as there was no strong incentive to standardize their appearance. They tended to be thinner than most coins, and almost all their decorative images were cut in low relief to enable the user to more easily stack and push them across the flat surface of a counting

Figure 1.3a and b Rechenmeister counter from Nuremberg, circa 1586–1635, from the author's personal collection. Images by Richard A. Otis.

board or cloth.[64] They often came in sets—also called casts or nests—of a hundred, but a set could be as small as thirty. Some inventories also list unusual set sizes such as 34, 39, or 89. This suggests that people occasionally lost counters but continued to use the sets as long as they remained large enough for ordinary calculations.[65] The most expensive counters were made of gold or silver; in 1535, John Longland, bishop of Lincoln, sent a dust box and a cast of presumably gold counters worth 40*li*. to Thomas Cromwell, Earl of Essex, while goldsmiths appraised a set of 40 silver counters at 20*s*. in 1600.[66] Cheaper stock counters, for everyday use by the general population, were made in brass, bronze, copper, latten, and lead; this was probably the kind the Exchequer spent a mere 7*d*. on when purchasing a purse full of counters in 1519.[67] Counters were produced in such large numbers that tens of thousands of counters have survived, though the metallic value of silver and gold counters was high enough that many were probably melted down and recast.[68]

Though counters were commonly used on a lined counting board, counters could be used alone if a board wasn't available. As Recorde explained in the *Grounde of Artes*, various methods of "accomptyinge by counters" fell into two general categories, "The one by lynes, and the other without lynes: in that *that* hath lynes, the lynes do stande for the order of places: and in *that* that hath no lynes, there muste be sette in theyr stede so many counters as shall nede, for eche lyne one, and they shall supllye the stede of the lynes."[69] Stacking counters to visually indicate implied lines made it possible for users to employ the system on any flat surface available, such as a shop counter, table, or even the ground.[70] While using counters "without lynes" appears to have been known throughout Europe, it was particularly common in France. Arithmetic textbooks explained how to form a vertical row of counters called a "tree of numeration" which functioned in the same way as lines.[71]

In England, counters were generally used with counting boards, and the phrase to "know the lines" meant to be familiar with counter-based arithmetic.[72] Counting boards could be cloths, or wooden tables with painted or chalked lines, rather than just a "board" as the name implies. While cloths could be laid out and taken up again after use, dedicated tables—such as the Exchequer's ten foot by five foot table—required permanent locations to store them.[73] Some households, businesses, and government offices maintained dedicated spaces, called counting houses, for accounting and counting board calculations.[74] As material objects, counting boards of all kinds were more ephemeral than metallic counters; the friction generated by sliding counters across these surfaces meant that owners had to regularly replace worn out cloths and repaint faded lines on tables, as evidenced by the new cloths "to cover the chequer" purchased for the Reading town hall in 1521, 1522, and 1523.[75] This was not a large economic barrier to use as, like counters, counting boards could vary widely in terms of cost and materials. In 1556, two counting tables, three counter boxes, and forty counters were considered expensive enough to be an appropriate New Year's gift for Queen Mary. By contrast a 1596 inventory of the Earl of Huntington's goods valued two cloths, including a compter cloth, at a mere 12d.[76]

There were numerous small variations in how counters might be laid out and subsequently manipulated, but Recorde's textbook describes three main methods of counter-based arithmetic in England: the standard, merchants', and auditors' forms. All three required users to have at least an implicit understanding of place value and zero. In the base-10 standard form, the lowest line on the counting board was the units line, and each subsequently higher line stood for a power of ten—ten, one hundred, one thousand, ten thousand, and so on for as many lines as the calculator needed. Three counters laid out on the units line stood for three but on the tens line they stood for thirty. If four counters were laid out on each of the hundreds and units lines then they would indicate the number 404 with (implicitly) zero in the tens place. The standard form also usually employed a subbase of 5 in the spaces between lines. Up to four counters could be placed on a line such as the units line and, when a fifth needed to be added, the counters would be removed from the

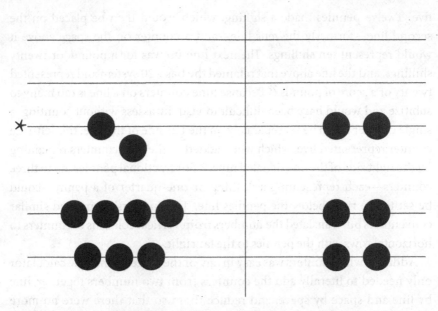

Figure 1.4 Counters laid out in base 10 standard form as 1543 and 2022. Image by the author.

units line and a single counter placed in the space above it signaling five. If there was already a counter in that space, they recognized they now had ten and needed to instead indicate that quantity on the tens line. They removed the counters from the units line and the counter from the space, replacing them with one counter on the tens line. Thus there were never more than four counters on any line—a quantity that could be recognized at a glance due to subitizing—or one counter in any space.

While the standard form was generically useful, the merchants' and auditors' forms were specifically designed to deal with the complexity of English currency and are likely to have been used by accountants on a day-to-day basis. Despite their eminent usefulness, Recorde was offended by the irregularities of the combined base-12 and base-20 forms; he included them briefly in his arithmetic textbook, but complained about "the order of commen castyng, wher in are bothe pennes, shyllynges, and poundes, procedynge by no grounded reason, but onely by a receaued fourme, and that dyersly of dyuers men: for marchauntes vse one fourme, and auditors an other."[77] In the merchants' form, the units line was replaced with a line for pennies and a counter on the space above it represented six pennies, not five. Twelve pennies made a shilling, which would then be placed on the second line—formerly the tens line—and a counter on the space above it would represent ten shillings. The next line up was for a pound, or twenty shillings, and the line above that retained the base-20 system and represented twenty or a score of pounds.[78] Because nine counters on a line is too many to subitize and would have been difficult to visually assess without counting, a single counter could be set separately on the left side of the line; this left-side counter represented five, which when added to the four counters remaining on the right side of the line signaled nine.[79] For fractional pennies, up to three counters—each representing a farthing or one-quarter of a penny—could be set to the right below the pennies line. The auditors' form used similar conventions but translated the numbers from vertical columns of counters to horizontal rows with the pennies to the far right.

Addition with counters was easy in any of the three forms. The calculator only needed to literally add the counters from two numbers together, line by line and space by space, and reduce them so that there were no more than the allowed number of counters on each line. Subtraction was also simple and involved the same process in reverse, occasionally borrowing counters from a higher line to subtract, say, 19 from 211. Calculators even

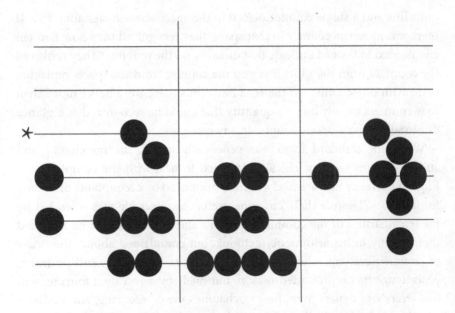

Figure 1.5 Addition in merchants' form, showing 35*li*.8*s*.2*d*. + 2*li*.2*s*.4*d*. = 37*li*.10*s*.6*d*. Image by the author.

had the choice of working from top to bottom or bottom to top (left to right or right to left in auditors' form), as the quantity of counters in each place value remained easily manipulatable throughout the process—in contrast to modern Arabic numeral subtraction, which requires setting down a number in pen or pencil then erasing or scratching it out to change it later. Multiplication and division were more difficult, as they required multiplying or dividing a number one line or space at a time and keeping track of which line or space the results should go on. In standard form, these place-value-based processes were markedly similar to today's processes for written Arabic-numeral-based multiplication and long division. But while it is theoretically possible to multiply and divide in merchants' and auditors' form—carefully accounting for the difference between the base 12 and base 20 lines—this does not appear to have been done in practice. Recorde flatly declared it impossible to do so and advised his reader, "in suche case, you must resort to your other artes," i.e., the base-10 standard form for counters or Arabic numeral arithmetic.[80]

The frontispiece of the 1543 edition of Recorde's *The Ground of Artes* shows a scene that might plausibly have resulted from mixing merchants' form counting board addition and subtraction with Arabic numeral division.[81] Four men crowd around a counting table, at least three of whom are directly involved in the calculations. The one on the left manipulates counters over the lines of a counting board that have either been temporarily chalked or permanently inscribed onto the table. Two men on the right stand over an Arabic numeral division problem that appears to be worked out directly on the table—possibly writing with chalk, but no slate—while the man at the back writes down the results of all their calculations on a piece of paper using Arabic numerals.[82] The image hints at Recorde's personal preferences both by the sole counting board user's spatial isolation and by the use of Arabic numerals for the date on the wall—the year of the book's publication, 1543.[83] However, his calculators' use of two different systems also suggests that both had a place in the realm of financial transactions; indeed, including counters in the image might have been necessary to advertise his book to his intended audience.

As Recorde's frontispiece illustrates, one of the advantages of calculations done with counters was that they lent themselves well to outside observation and, in the realm of financial transactions, outside audit. Customers could watch shopkeepers calculate their bills and merchants could watch custom officials calculate their duty payments in order to ensure the accuracy of

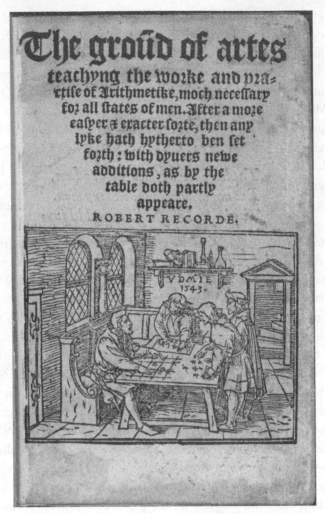

Figure 1.6 Woodcut from the title page of the 1543 Robert Recorde's *The Ground of Artes*. © British Library Board. All Rights Reserved/Bridgeman Images.

the calculations. Sheriffs bringing their accounts before the Exchequer had their income and expenditures "read and cast up in open Court, before the *Treasurer & Barons* of the said Court, in the presence of the *Treasurers Remembrancer, Clerk of the Pipe, Controller,* & other Officers whom it shall or may concern; and two *Auditors* at the least for casting up the said accounts."[84] Calculating and double-checking the calculations of these accounts was both a mathematical activity and a public performance before an audience of interested—or at least legally invested—parties.

This performative aspect of counters, combined with their ability to take imprints in the same fashion as coins and medals, resulted in a form of conspicuous numeration during the sixteenth and seventeenth centuries. While high medieval counters were minted with generic images like crosses, keys, rakes, or scales, even the most generic early modern counters generally had more complex designs.[85] Counters most frequently spoke to their owners' skill with the physical act of calculation, bearing legends such as "cast up correctly, cast up, and you will find the amount," "he who casts up correctly will find his account," and "diligent accounting leads to correctness"; the common Rechenmeister style of Nuremberg counters paired such legends with the letters of the alphabet and an image of a man performing counting board arithmetic.[86] These images and legends, asserting the calculators' skill, would then be prominently displayed to onlookers through every act of calculation. Counters bearing the image of the current monarch—and performatively asserting loyalty to that monarch—were also perennial favorites and had wide appeal for people of all social statuses.[87] Nicholas Hilliard, engraver to Elizabeth I and James VI/I, issued counter sets with images of members of the royal family, "for reckoning and for play," and obtained a presumably lucrative twelve-year monopoly over such images beginning in 1617.[88]

In addition to the generically marketable images of calculations and monarchs, many sixteenth- and early seventeenth-century nobles and elite government officeholders commissioned personalized counters to testify publicly to their owners' political and religious loyalties. Such counters served as status symbols, particularly for men who could boast of their public offices on counters they might use to calculate with during the course of their duties. Other people commissioned counters for their own private use in accounting, proclaiming their personal titles rather than their public positions.[89] The use of personalized counters flourished among English elites until the civil wars in the middle of the seventeenth century. After the Restoration, a new generation of nobles and public officeholders continued to strike medals with their mottos and images but no longer showed an interest in commissioning their own, personalized counters.

Although Nuremberg counters were regularly produced for another century, with extant counters bearing the image of every British monarch from Charles II to George II, the quality of these mass-produced counters declined. They were used less and less for financial calculations and instead retained only as tokens for scoring games.[90] Of a set of 1,212 counters dredged primarily from the river Thames in London, 92% dated from 1300 to 1650, at which point there was a sharp drop-off in number; there were only six post-1750 counters.[91] The final, 1699 edition of Recorde's arithmetic textbook dropped its section on counters, and Baily's 1740 *An Universal Etymological English Dictionary* defined a counter as "A piece of Brass or other metal with a stamp on it, formerly used in counting, but now in Playing at Cards."[92] That said, a 1753 French arithmetic textbook asserted that counters were still widely used, including by government clerks, though they were more often used by women than men, presumably due to differences in access to education or material resources.[93] While intended to discuss French numeracy, the statement could have equally applied to England, where the Exchequer continued to use its counting table until the early 1780s. Its use was discontinued only around 1785, not long after the 1783 act that reformed the Exchequer and (eventually) abolished the use of tally sticks.[94] In France, the counting board lasted a few years longer but use of counters for arithmetic "ceased entirely" during the French Revolution.[95] Nuremberg counters continued to be produced into the nineteenth century, but as game tokens for tallying scores rather than as the foundation of a symbolic system used for arithmetic.

Roman and Arabic Numerals: Written Symbolic Systems

Object-based symbolic systems like tally sticks and counters often had a written component, but only three systems required the user to possess the tools and skills of literacy. The simplest system involved writing number words using the same alphabetic symbols as any other English word. The other two systems involved special symbols to indicate numbers. The older of the two was the additive system of Roman numerals, which was adopted throughout Europe during the period of the Roman Empire and remained the primary written symbolic system of medieval and sixteenth-century England.[96] During the late sixteenth and seventeenth centuries, however, Roman numerals were gradually replaced in most contexts by the newer positional system of Arabic numerals.[97] This symbolic system was first developed in India and subsequently adopted in the Islamic Middle East and Iberia before being introduced into Christian Europe. The widespread adoption of Arabic numerals in England is particularly important for the study of numeracy and arithmetic, not just because it remains one of the most dominant symbolic systems in the twenty-first century, but also because it enabled people to adopt written calculation methods based on Arabic numerals.

Number words formed an essential part of the English language and were learned alongside other English words. As a consequence, written English number words required the same skills as writing any other English word: knowledge of the Latin alphabetic symbols and the ability to use the tools of literacy. These tools were often pen and paper but could also include writing with a stylus on erasable wax tablets, engraving, embroidery, or other media. Such written number words were cumbersome in length but did not require the literate user to memorize any new symbols or syntactical rules for constructing numbers. This system thus formed a subcomponent of, and was taught in conjunction with, the ability to read and write in English. Written number words became increasingly familiar to the English population as literacy rates rose throughout the sixteenth and seventeenth century. They were, however, confined to the literate part of the English population.

Roman numerals were a simple and intuitive symbolic system that likely developed from object-based tally systems; the first symbols—I, II, III, and IIII, which can also be written as IV—are visually equivalent to a simple tally.[98] These symbols could be easily taught, so long as the student already knew the words for numbers up to a thousand and had a basic grasp of mental addition and subtraction. There were only seven symbols to memorize, all of which had analogs in the English alphabet, and the system's additive structure required little explanation. Brinsley spent a mere two paragraphs on this lesson, explaining to his readers that I stood for one, V for five, X for ten, L for fifty, C for a hundred, D for five hundred, and M for a thousand. As for how these letters were to be used, he noted that

> any number set after a greater, or after the same number, doth adde so many mo [sic], as the value of that later number is. As, I. set after X. thus, XI. doth make eleuen. XV. fifteen. XX. twentie. But being set before, they doe take away so many as they are: as I. before X thus, IX. nine. . . . And thus much shortly for numbring by letters.[99]

However, despite the additive construction of Roman numerals, they in no other way facilitated arithmetical calculations. This has been a source of considerable confusion to scholars, who assumed Roman-numeral arithmetic must have been common and have subsequently invented convoluted methods for multiplying and dividing with them.[100] However, early modern men and women would not have attempted to multiply with Roman numerals. Those who wished to add multiple numbers, multiply, divide, or extract roots would instead turn to a different symbolic system more suitable for performing their calculations. The counting board was ideal for these purposes.[101] In its standard form, its base 10 lines and subbase 5 spaces neatly aligned with the construction of Roman numerals. Once calculations were complete, a user could easily transcode the results into Roman numerals. Early modern men and women who used Roman numerals thus did so in the context of multiplistic numeracy, regularly shifting between symbolic systems as needed.

The place-value system of Arabic numerals was new compared to Roman numerals, but by the sixteenth century it had a relatively long history in England. Although initially developed in India, the symbols became known as Arabic numerals because Europeans first became aware of them through Arabic scholarship in the Islamic world. Arabic numerals entered Europe through Iberia, where the western variant of these numerals were known as ghubār (dust) numerals, likely in reference to their use on dust boards.[102] Scholars such as Gerbert of Aurillac, the future Pope Sylvester II, brought Arabic numerals from Iberia to Italy during the late tenth century, where a rotated version of these numerals was used to mark the stones of a modified "Gerbertian" abacus that never achieved widespread use. However, by the twelfth century, a variety of Arabic works were being regularly translated into Latin by scholars in Toledo, to the point that Christian scholars sometimes called the system Toledan numerals.[103]

Their earliest known appearance in England dates to 1130, when Adelard of Bath translated Persian scholar Muhammad ibn Mūsā al-Khwārizmī's treatise on Arabic numeral arithmetic into Latin.[104] Around the same time, one of Adelard's students—known only as "H. Ocreatus"—attempted to merge the Roman numerals I through IX with Arabic numerals' positionality and explicit zero to create a hybrid that combined the traits of both systems. Positions were separated with dots, a feature that would later be reinvented to separate different denominations of currency.[105] As with earlier attempt to introduce Arabic numerals to Italy over a century earlier, neither Arabic numerals nor Ocreatus's hybrid system gained much traction among users of the existing English symbolic systems; Adelard himself continued to use Roman numerals for the rest of his life. By the fourteenth century, versions of Arabic numerals began to appear in specialized mathematical manuscripts— including astronomical tables, nativity calendars, and psalter calendars.

However, it would be another two centuries before vernacular, printed arithmetic textbooks helped stabilize the appearance of Arabic numerals into the symbols 0 through 9—still used by Europeans and residents of their former colonies today—and made information on Arabic numeral arithmetic available to the wider English population.[106]

When early modern English men and women did begin to adopt Arabic numerals, they did so in a piecemeal fashion that was consistent with multiplistic numeracy, picking and choosing the contexts where Arabic numerals would be most helpful rather than completely replacing one or more of their existing systems. There is a clear pattern of Arabic numerals initially being adopted for use in a single context—specifically, to record the number of years since the incarnation of Christ, e.g., 1509.[107] During the late sixteenth and seventeenth centuries, people subsequently broadened their Arabic numeral use to include an increasing number of diverse contexts. Although there was considerable variation in the timing of individuals' first use of Arabic numerals to record information, there was a fairly consistent pattern in the order of contexts where these numerals appeared in an individual's accounts: first in the year of the incarnation in engravings and written documents, followed by other aspects of dating, quantities and monetary values, and finally the summation of accounts.

While fourteenth-century use of Arabic numerals seems mostly confined to mathematical elites, by the fifteenth century more English men and women were employing Arabic numerals for the year of the incarnation, particularly in engravings on buildings. The year of the object's creation, as measured from the incarnation, was the most typical number engraved on an object, and it is likely that the inherently shorter length of the year, as recorded in Arabic rather than Roman numerals, appealed to the objects' creators. The British Museum has astrolabes from 1326 and 1342 inscribed with Arabic numeral years of incarnation, while as early as 1424, the seal of Fountains Abbey bore the Arabic numeral year 1410.[108] Later in the fifteenth century, people began to use Arabic numerals to record the year on wooden gates and doors, stone walls, pavement, and decorative shields. By the turn of the sixteenth century, Arabic numerals could be found recording the year on a wide variety of material objects, including tabernacles, wood-cuts, roof beams, tombs, brasses, bells, cups, windows, and everyday mathematical objects like counters; the earliest counters with dated legends come from this period and almost always used Arabic numerals.[109]

By the early to middle years of the sixteenth century, English men and women expanded their use of Arabic numerals for the year of the incarnation into their written accounts and letters. With the exception of merchants and tradesmen—who had more contact with Continental merchants and a greater financial incentive to adopt their accounting practices—English account-keepers generally retained Roman numerals as the primary method for recording quantities, monetary costs, and other kinds of dates, including the regnal year. The churchwardens' accounts of St. Michael, Spurriergate, York, first contained an Arabic numeral year of the incarnation in 1547, while four Peterborough churchwardens recorded their receipts in Roman numerals from the "feast of S. Mychell in Anno 1554 vntill the Feast of S. michell in Anno 1557."[110] In the parish of Prestbury, the register book listed Arabic numeral years beginning in 1560, but did not use Arabic numerals for any other purpose until 1611–2.[111] Even in 1627–8, the churchwarden Nicholas Goore in Walton-on-the-Hill, Lancashire, used Arabic numerals for the year and Roman numerals in all the remaining quantities, save one.[112]

Although early modern English men and women initially saw no need to use Arabic numerals for recording anything other than the year of the incarnation, they eventually began to extend their use of Arabic numerals to other aspects of recording dates. In the 1530s, Honor Plantagenet, Lady Lisle, wife of the Lord Deputy of Calais, received letters employing both Roman and Arabic numerals for days of the month. Henry Higford, the clerk of the corporation of Stratford-Upon-Avon in the late 1560s, consistently recorded the year of the incarnation in Arabic numerals, while varying between Arabic and Roman numerals for the day of the month.[113] The parish registers of Prestbury recorded christenings performed on "29th Marcij" and "vjto Aprilis" in 1623, while the Peterborough feoffees used a similar mixture of Arabic and Roman numerals for days of the month throughout the 1610s and 1620s before dropping the use of Roman numeral dates.[114] Account-keepers also began to use Arabic numerals to record quantities and prices, as well as dates. In his parish register, Thomas Hassall of Amwell used Roman numerals for the days of the month until 1611, and used a mixture of Roman and Arabic numerals until 1636. As early as 1603, however, he followed the convention of the London Bills of Mortality in noting the total dead in his parish in Arabic numerals: "Buried in all this yeare 41" and "of the plage 19."[115] A 1615 list of Adventurers for Virginia in the first Wycombe ledger book list investments of "10s.," "5s.," "40s.," "13s. 4d.," and "20s." Churchwarden Nicholas Goore noted that he had "paid the glazier for worke done at the church/for Fourteen Foote of new glass at 6*d*/the foote" in 1627–8.[116]

As account-keepers began to employ the Arabic numerals in a wider variety of places, overall they used the Arabic and Roman numeral systems side by side and interchangeably. The church records of St. Andrew Hubbard include a 1534 will in which hatmaker William Harber bequeathed "in 5 Friday every 5d. sterling to v poor, needy persons in the honour and worship of the v wounds of Our Lord Jesus Christ."[117] The Prescot churchwardens' accounts record a payment for "8 ashlers" immediately before a payment for "ij daies worke in layinge the ashlers," while the Peterborough feoffees' accounts note "all soe xxs. of the same Some is in like Mannour to bee oute being parte of the 60*li.* of Mrs. Swinscotes gift."[118] Fluent use of multiple symbolic systems even occasionally manifested itself in quantities that can best be described as "mixed," in that they use both Arabic and Roman numerals in the same number. This sometimes occurred in years such as "anno M quingentesimo 24" and "Mij. 63" or monetary values such as "xxvii*s* 8*d*" and "ii*s* 6*d*."[119] When recording quantities or prices, however, the different symbolic systems were more commonly used in separate numbers, though they might be employed in the same sentence or entry. Mixed numbers were never used frequently enough to create a hybrid system and ceased to appear in records after about 1650.[120]

Around the turn of the seventeenth century, people also began to use Arabic numerals for the final type of numbers that appeared in their accounts: valuation columns and sums. Peter Wardley and Pauline White's collaborative study of probate inventories specifically focused on this context, "ignoring numbers in the text, dates and quantities of objects" where Arabic numerals were already in use. Although they briefly noted "the likelihood of binumeracy on the part of some" people who used both Roman and Arabic numerals in their probate inventories, they were primarily interested in what they called complete mastery of Arabic numerals—by which they meant the complete abandonment of Roman numerals.[121] They found that most people in their sample adopted Arabic numerals for use in valuation columns between 1590 and 1650, with the main transition period beginning in the 1620s and 1630s and continuing through to 1650. This corresponds well with Wardley's earlier research on Bristol and West Cornwall, where he found a slightly earlier adoption period of 1570 to 1630, and J. M. Pullan's examination of the Bristol archives, which showed a transition period of 1635-40.[122] The Norfolk probate inventories showed some of the earliest uses of

Arabic numeral sums, in 1584, while most of the places Wardley and White surveyed had earliest adoption dates between 1607 and 1612. However, this was by no means a universal trend. Accountants in Bermondsey and Rotherhithe used Roman numeral sums until the end of the seventeenth century, and individuals generally continued to use Roman numerals alongside Arabic numerals in most of the surveyed areas.[123]

While it is a relatively straightforward—if time consuming—task to track the use of Arabic numerals in still-extant written records, it is far more difficult to determine when people began using Arabic numerals for calculation as opposed to recording numerical information. Although it is not always possible to determine what symbolic system was used for any particular account, some general trends can be seen. Most fifteenth- and early sixteenth-century accounts, especially those written entirely in Roman numerals, were most likely written by someone who used a counting board for their calculations. The late sixteenth and seventeenth centuries then saw a proliferation of arithmetic textbooks intended to educate people in Arabic numeral arithmetic. Late seventeenth- and eighteenth-century accounts, or earlier accounts written entirely in Arabic numerals, were more likely written by someone who had adopted Arabic numerals for both recording and calculation. In between, many people probably transcoded between symbolic systems as needed and desired, choosing the system that seemed easiest or most appropriate in any particular context.

Given the multiplicity of early modern numeracies, people's transition from written Roman numerals and counting board arithmetic to written Arabic numerals and Arabic numeral arithmetic was not a simple process of immediate or complete replacement but rather a series of context-specific decisions that led to a gradual transition from one collection of systems to another. The slow-growing popularity of Arabic numeral arithmetic was reflected in the changing terminology they used to distinguish it from other forms of arithmetic over the late medieval and early modern periods. From the thirteenth century, Arabic numeral arithmetic had been known as "algorism," after al-Khwārizmī whose translated works introduced it to England. By 1530, it was also known as "cyphering," to reflect its positional dependence on the explicit concept of the cypher, or zero.[124] John Palsgrave included both words in his 1530 French-English dictionary, translating sample phrases such as "I Cyfer[,] I acompt or reken by algorism."[125] However in the seventeenth century, terms intended to distinguish arithmetic by Arabic numerals from arithmetic by counters began to drop out of use, and "arithmetic" gradually became synonymous with Arabic numeral arithmetic. By 1701, Scottish physician and mathematician John Arbuthnot could declare it "would go near to ruine the Trade of the Nation, were the easy practice of *Arithmetick* abolished: for example, were the Merchants and Tradesmen oblig'd to make use of no other than the *Roman* way of notation by Letters" in conjunction with counters, tally sticks, or other symbolic systems.

———

Early modern English men and women employed a variety of symbolic systems to convey numerical information, transcoding freely among performative, object-based, and written systems. Oral systems were the most universal and served as a basis for acquiring a familiarity with other symbolic systems. Finger-counting was a widespread aid to mental arithmetic, but other gestural systems were considered curiosities and were promoted as methods for secret communication. More common were object-based systems such as tally sticks and counters. These objects could be invested with both numerical and non-numerical information, but only the former could be transcoded into other symbolic systems. Both performative and object-based systems were employable by those with varying degrees of literacy and were institutionalized for use in government courts such as the Exchequer.

Number words and Roman numerals were the dominant written symbols at the beginning of the early modern period and were employed in conjunction with counters and counting boards. Over the course of the sixteenth century, people adopted Arabic numerals in a context-specific fashion consistent with multiplistic numeracy. This began with dates and gradually expanded to include quantities and sums during the first half of the seventeenth century. After the Restoration, the population increasingly abandoned Roman numerals, counters, and counting boards for Arabic numerals and written arithmetic. While people still used multiple symbolic systems at the end of the seventeenth century—and, indeed, people in England and its former colonies still operate in a context of multiplistic numeracy and use objects in support of their calculations today[126]—Arabic numerals had become the most dominant symbolic system in England and the standard by which eighteenth-century numeracy would be judged.[127]

2

"Finding Out False Reckonings":
Trust and the Function of Numbers

In 1635, Gervase Markham published a treatise on husbandry and farm management in which he argued a servant's trustworthiness was far more important than his or her skill with letters. However, the more Markham discussed
his servants' learning, the more he shifted his discussion of trust into terms of
accounting and numbers. He conflated numeracy with literacy as he argued:

> there is more trust in an honest scoure chaulkt on a Trencher, then in a cun-
> ning written scrowle, how well so ever painted on the best Parchment....
> I had rather be my Mans *Amanuensis* to register his Truthes, then a
> Witnesse of his Learning in finding out false Reckonings. And there is
> more Benefit in simple and single Numeration in Chaulke, then in double
> Multiplication, though in never so faire an hand written.[1]

He was not completely opposed to writing—in particular, he believed "as
touching the Master of the Family himselfe, learning can be no Burthern."
But he refused to put too much stock in it either, as writing could easily be
"falsified and corrupted," while a servant's integrity "if it be sound will hardly
be shaken."[2]

Markham not only made trust judgments based on the actions of his
servants, but he also extended those judgments to cover the symbolic systems
those servants used to record information. Chalked scores were "honest," a
straightforward and a trustworthy system that was not prone to error. By
contrast, he denigrated written systems as "cunning"—playing off the word's
dual definition of both being learned and covertly deceitful—and argued this
deceitfulness was an attribute that writing retained, regardless of whether
or not it was presented in an aesthetically pleasing manner.[3] However his
examples suggest that he was less mistrustful of writing in general and more
wary of a new type of written calculation—"false reckoning" and "double
Multiplication" in writing with Arabic numerals.

By the Numbers. Jessica Marie Otis, Oxford University Press. © Oxford University Press 2024.
DOI: 10.1093/oso/9780197608777.003.0003

In the context of early modern numeracy, symbols formed a vital intermediary between people creating and sharing numerical information.[4] Trusting each other was insufficient if the system itself was untrustworthy. Symbolic systems were differentiated from each other by their material characteristics, which helped form the basis for people's trust judgments and influenced people's decisions about whether or not to trust the system to perform a specific mathematical function. Trust is a key factor in explaining early modern patterns of symbol use as well as Arabic numeral adoption during the sixteenth and seventeenth centuries.

Recording versus Calculation

In describing his ideal servant, Markham made judgments about the trustworthiness of different symbolic systems, but he also expressed a desire for his servants to limit themselves to recording numerical data—"simple and single Numeration"[5]—rather than calculating with it. While there does not seem to have been a strong conceptual difference between recording and calculating in early modern writings about mathematics, Markham's complaint highlights an important practical distinction between recording quantitative information with numerical symbols and performing arithmetical calculations with the assistance of numerical symbols. Early modern mathematicians divided arithmetic into several "parts," the first of which was always numeration or knowledge of the number symbols themselves. Indeed, numeration requires an understanding of the cardinality principle—that the last number in a count gives the quantity of objects counted—and a built-in series of implicit calculations: specifically, adding one to a given quantity. However, subsequent parts of arithmetic—addition, subtraction, multiplication, division, and then various rules of three—all involved more explicit calculation and were effectively distinct skills from numeration.

More fundamentally, early modern men and women favored different material characteristics in the symbolic systems they used for recording and calculating. Creating a record, particularly a permanent one, required fixing a numerical symbol so that it could not be easily altered. People who sought to record numerical information ascribed greater trustworthiness to systems with more permanent and less manipulable symbols. They even developed strategies to further decrease the manipulability of such symbols, thus increasing the trustworthiness of the system for conveying information across time and space. By contrast, people who performed numerical calculations favored manipulable symbols. The very act of calculating transformed two or more numbers into a single, numerical answer. There was thus an inherent tension between recording and calculation, with their conflicting needs for permanence versus manipulability. As a result, early modern men and women tended to choose different symbolic systems for one function or the other, depending on the system's material characteristics and affordances.

This practical distinction between recording systems and calculating systems is particularly vital for understanding patterns of Arabic numeral adoption in early modern England. Generations of scholars have focused on the adoption of Arabic numerals for calculation, believing this was their primary appeal. And for scholars who come from a culture steeped in Arabic numeral–based mathematics, it is easy to retrospectively assume that as soon as the barrier of illiteracy could be overcome, "the western world would at once adopt the new numerals . . . which were so much superior to anything that had been in use in Christian Europe."[6] Indeed, previous historians have wondered at the "stubborn resistance to the new numerals"[7] and denigrated those "few reckoners who went on using the old, obsolete methods."[8] One noted the existence of a viable alternative in the counting board, and the confusion arising from an explicit zero, as the main "intellectual obstacle" to the widespread adoption of Arabic numerals.[9] For these scholars, the Arabic numeral system's primary rival was the counting board, and their primary interest lay in its adoption for calculation.

In general, however, the men and women of early modern England first adopted Arabic numerals not for calculation but for recording. They didn't even use them for recording all types of numerical information but specifically for shortening the length of large numbers in dates. Like Markham, they mistrusted Arabic numerals for more sensitive financial recording. And even as people increasingly adopted them for recording, the multiplicity of early modern symbolic systems made it common for people to use Arabic numerals only for recording, continuing to rely on the tried, true, and trustworthy counting board for calculation.

From an early modern viewpoint, Arabic numerals were not the most trustworthy of systems. But they did have a functional advantage over other symbolic systems: Arabic numerals enabled the unification of permanent recording and calculation in a single system. Early modern men and women could, and did, use Arabic numerals for solely one function or the other.

However, the potential for increased efficiency encouraged people to adopt the unification of both functions, even in the face of Arabic numerals' other limitations. During the last decades of the sixteenth century and the first half of the seventeenth century, Arabic numerals' enhanced functionality increasingly—but not universally—trumped the security of other symbolic systems.

Permanence and Trustworthy Records

The trustworthiness of a system for recording could be judged by its ability to create a permanent record of numerical information that persisted across space and time, particularly one resistant to later alteration. Not all alterations were made with malicious intent, but systems that allowed such alterations were more vulnerable to fraud. Of the seven most common early modern symbolic systems for numbers, four were suitable for creating a permanent record. The performative systems—spoken words and gestures— were necessarily ephemeral; unless they were subsequently translated into written form or engraved on a material object, the numbers existed only· as long as someone was speaking or gesturing. The object-based system of counters and counting boards was similarly ill suited for creating permanent records because counters were designed to be easily moved around a board and could not be fixed into a specific, unalterable location. By contrast, the object-based system of tally sticks and the three written systems of English number words, Roman numerals, and Arabic numerals all created permanent records with varying degrees of resistance to later alteration.

Of all the early modern symbolic systems, the tally stick was considered the most secure and trustworthy, with its split format specifically designed to create a permanent record that could not be changed without leaving obvious signs of tampering. Given the scarcity of coins in early modern England, and the widespread use of reciprocal credit to ensure the smooth functioning of the economy, the trustworthiness of recorded credit transactions was vitally important.[10] The shape and grain of each tally stick was unique, and the security of its notches was ensured by splitting the stick into two unequal parts, which were then divided between the creditor and debtor. Any attempt to alter or add to the notches on either part would be immediately apparent when the two were brought back together to check the account. A 1589 reprint of a medieval book on husbandry advised readers to "not sell, buy, receyue, nor deliuer any thing except by tally, or by good testimonie," thereby rating the tally stick equivalent to the word of a known and trustworthy man.[11] In 1622, a preacher described the practice of "Bakers and Brewers (& such *Tradesmen* as vse to deliuer their commodities to their Customers *vpon Tale*)" as using a split tally "to keepe euen reckonings: So that if either goe about to deceiue other, they neede but to drawe our *Their side of the Tally*, and *That* will soone discouer the Truth."[12] The tally stick's testimony could therefore be considered more trustworthy than that of tradesmen or their customers. The two parties might sometimes agree on the number recorded by the tally but not on its real-world meaning. For example, a fourteenth-century Lord Chancellor of England complained he "cannot tell whether [the notches] refer to bullocks or to cows or to what else . . . so we hold this [tally] to be no deed which a man must answer."[13] However, only when all or part of a tally was lost or destroyed—invalidating its security measures and forcing creditor and debtor to rely instead on each other's verbal assurances—could the recorded numbers themselves be questioned.[14]

Because of this trustworthiness, the split tally stick was particularly associated with the Exchequer as "a very antient and most certain way of avoiding all cozenage in the Kings Revenue." While people could be very motivated to avoid paying debts to the government, the tally stick ensured that any "corruption . . . is easily and soon discovered, and the Offender severely punished by Fine and imprisonment."[15] When one late medieval creditor added 60s. to the notches on his Exchequer tally stick, his deceit was promptly discovered, and he was sentenced to a year and a day in prison.[16] Even after rising literacy rates had enabled the wider adoption of Arabic numerals in the late seventeenth century, authors were still extolling the virtues and security of the tally compared to written alternatives:

This Antient way of striking of *Tallies* hath been found, by long experience, to be absolutely the best way that ever was invented, for it is Morally impossible so to Falsifie or Counterfeit a *Tally*, but that upon rejoyning it with the *Counterfoyl*, it will be obvious to every Eye, either in the *Notches*, or in the *Cleaving*, in the *Longitude, Latitude*, Natural growth or shape of the *Counterfoyl*; whereas *Acquittances* in Writing cannot be so done, but that they may be *Counterfeited* by skilful Penmen, and that so exactly, as that he who wrote the Original, shall not be able to know his own hand from the *Counterfeit*, as hath been frequently seen in all the *Courts* of *Westminster*.[17]

The split tally stick was designed for secure transactions and fraud detection, and there appears to have been little to no early modern concern about attempts to counterfeit or alter split tallies.

While this system was considered unimpeachable, tally sticks' main practical limitation in the realm of finance was the sometimes problematic credit of their issuer. When Samuel Pepys was unable to convince a goldsmith to trade a tally for money, no one doubted the veracity of the tally stick, only the ability of its issuer to make good upon the debt.[18] A few years later, an alcoholic found herself unable to obtain further money from the tally shops because her husband "told them if they trusted her any more he would not pay them."[19] His forthright refusal to pay his wife's future tallies, much like Charles II's suspension of regular Exchequer payments from 1672 to 1677, laid bare some of the problem creditors might face in collecting debts via tally.

By the middle of the seventeenth century, tallies still held legal force with respect to government debts but creditors ran into more legal challenges when using private tallies. In 1641, lawyer and politician Thomas Wentworth argued that debts recorded on tallies were promises, like a man's given word, but they did not have the same legal force as a sealed, written contract unless they were issued by the government. Failure to honor them could be prosecuted as breach of promise, and damages recovered that way, but the executor of a debtor's estate was not actually obligated to pay the original debt.[20] A 1648 treatise similarly warned that obligations made "on a Tally, peece of wood, or any other thing but paper or parchment, albeit it be sealed and delivered, yet it is voyd." In 1651, another treatise specifically flagged the written component of tallies as lacking legal force by arguing a tally could not be considered the same thing as an "Obligation in writing . . . because letters written in wood may be easily raced out, and altered, it was resolved unfit to allow and open so apparent a way to deceit."[21]

While obligations written on paper or parchment had the most legal force, both government and private tallies remained widely trusted among the wider English population. Tradesmen and alehouse-keepers continued to run tabs with tallies, and miners used tallies to keep track of their work and ensure they received correct payment from company officers.[22] Much like modern gift certificates, tallies were effectively equivalent to money. Samuel Pepys, in his diary, fretted equally over being robbed of tally sticks and being robbed of specie. When he fled during the Great Fire of London, the survival of his tally sticks was one of his main concerns: he "got my bags of gold into my office ready to carry away, and my chief papers of accounts also there, and my tallies into a box by themselfs."[23] At the end of the seventeenth century, one Scottish reformer even proposed to formalize this equivalency and issue tallies designed to be used as "passable Money."[24] Although rising literacy rates made the three written systems increasingly available alternatives, tallies' trustworthiness, versatility, and transferability ensured their continued popularity into the eighteenth century.

Of the three written systems, only writing out the English number words came close to matching the permanence and ascribed trustworthiness of tally sticks. The main ways the integrity of these systems could be compromised were the alteration of existing symbols or the addition of new symbols. In general, the alteration of a few letters within a word could not change the meaning of the word itself without leaving evidence of tampering, allowing exceptions for very skilled forgers as well as the very similarly spelled "seven" and "eleven. " While extra letters and words could more easily be added to a number—such as turning "six" into "sixteen" or "forty-three" into "two hundred forty-three"—this required sufficient space on the page around the original number. Such alteration could easily be prevented by embedding the number in a line of text. The expansiveness of written number words was thus both a strength and a weakness, preventing the alteration of numbers while requiring significantly more space than the other written symbolic systems, particularly for large quantities.

Roman numerals were also a relatively difficult symbolic system to alter, though they were not as secure as tally sticks and number words. I, V, X, C, D, and M were distinct symbols that could not be easily exchanged for one another without leaving visible proof on the page. The main place the Roman numeral system was vulnerable to tampering was at the beginning and end of each written number. For instance, XI could easily be transformed into XXI or CXI by the addition of a numeral at beginning of the number. Similarly, it could become XII, XIII, XIIII, or XIV by the addition of more I's or a V at the end of the number. While many early modern accounts used the subtraction-based IV to indicate the number four, others used IIII to indicate the same thing.[25] The use of one or the other seems to have been a matter of personal preference, as opposed to standardized usage, providing the opportunity for someone to alter the right side of the number in a more-or-less space-intensive manner as desired.

When Roman numerals were used within a line of written text, they were generally safe from this type of tampering. As with number words, adding a numeral would crowd the text immediately before or after it and be noticeable at a glance. However, most early modern accounts were set up in two columns, with the left column containing the description of an expense in textual form and the right column containing only the cost to be recorded, often with a sum of all the costs at the bottom of the column. The free-floating numbers of the right-hand valuation column were both high-value targets for fraud and particularly vulnerable to tampering by the addition of extra numerals.

Account keepers had three main strategies for minimizing this risk, all of which involved spatially isolating the otherwise free-floating numbers: placement, the addition of non-numeric symbols, and the modification of the existing symbol set. Placement generally entailed justification of numbers, which made it impossible to add extra symbols to the one end of the number without destroying the clean vertical line of the number column. Left justification was more common than right justification, both because English men and women wrote left-to-right and because adding a symbol to the left end of the number would create a larger number than adding a symbol to the right end. Writers could also add horizontal and vertical lines, along with brackets, to spatially locate the beginning or end of numbers in a similar fashion. The addition of non-numeric symbols generally involved periods or colons placed before and after numbers in order to isolate the number from any attempted addition, or the use of monetary denomination symbols such as £, s., and d. to one side of the number. Finally, the symbol J could be substituted for the terminal I in a number—i.e., XVIJ instead of XVII—and thereby prevent any subsequent addition to the right of the number The terminal J probably originated in southern Italy around the year 900, though a few classical texts used an upside-down J in a similar fashion.[26] Most account keepers did not rely on a single strategy, but mixed and matched among them. The following account of Prince Arthur's funeral expenses in 1502, for example, used a combination of left justification, horizontal lines and brackets, periods, denomination symbols, and terminal Js to constrain the otherwise free-floating Roman numerals.

Of all the potentially permanent systems, Arabic numerals were the most vulnerable to tampering. The place-value system allowed a person to add any of the ten Arabic numerals to the end of a number, and all but the zero to the beginning of a number, providing a greater range of possible alterations. However, most of the strategies that people used to minimize risk for Roman numerals could also be used with Arabic numerals. Placement and the addition of non-numeric symbols did not rely on any particular characteristic

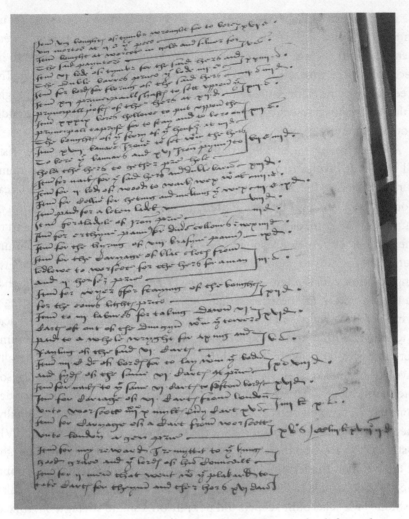

Figure 2.1 Roman numeral account containing Prince Arthur's funeral expenses. LC2/1, f. 20r. Used by permission of The National Archives of the UK (TNA).

of Roman numerals. Only the practice of using a terminal J was not directly translatable to Arabic numerals.

More problematically, Arabic numerals were the only written system where the symbols themselves were easy to alter. The Arabic numeral 1 could be invisibly transformed into a 4 or a 7. With a little more effort, it could be transformed into most of the other numerals as well. Similarly, 0, 6, and

9 could be interchanged with the addition of an ascending or descending mark. These and other alterations were possible in both manuscript and print documents, depending a person's handwriting or the print quality of a text. The changes did not even have to be deliberate. Sloppily written 6s or 0s can easily be mistaken for one another, while 1s and 2s are extremely difficult to differentiate in texts with low print quality, including many of the surviving London Bills of Mortality.[27]

Given this greater vulnerability to tampering, early adopters of the Arabic numeral system often preferred to retain more secure and trust- worthy English number words and Roman numerals for sections of their ac- counts that were at higher risk of tampering. People initially adopted Arabic numerals for the year of the incarnation, a date that was relatively safe from alteration as it would be evident if someone added another numeral to the beginning or end of the year; furthermore, the year was often clear from context. They subsequently began to use Arabic numerals for all aspects of date-keeping and for numbers protected within lines of text, such as in the left-hand column of account books. However, many people—such as the Elizabethan merchant John Isham and the Caroline bailiff of Peterborough— preferred to retain Roman numerals for the free-floating numbers in the right-hand column of account books, even after Arabic numerals had be- come their system of choice for all other numbers.[28] A 1348 edict by the University of Padua required the prices on books for sale to be written "'non per cifras, sed per literas clara' (not by figures but by clear letters)." A 1590 advisor to the Cannock ironworks in England agreed; other things might be written with Arabic numerals, but "Prises, for to avoide mistaking, are better written in letters."[29]

When Arabic numerals were employed to record the price of an item in the right-hand column of an account, the numbers were often further spa- tially isolated into three separate columns—one each for pounds, shillings, and pence. Another common strategy among account-keepers was to record the price with six or more digits, even when such length was unnecessary. For example, in an Exchequer certificate of revenue from 1679, 4li. 8s. was recorded as 0004: 08: 00. The additional zeros prevented minor alterations to entries that might transform a single penny into 10 or 11 pence. However, this did nothing to prevent alteration of the numerals themselves. This docu- ment has at least three altered numerals, of varying degrees of visibility to the casual reader. An entry for 00s. has been clearly altered to 08s., as the second circle extends awkwardly upward from the first, in comparison to the tilted, sideways 8s. visible in other numbers. In an entry for 01s., the awkwardly straight right-hand side of the 0 indicates it was probably originally a 1, while the slight difference between the 1 and other 1s on the page as well as the discoloration of the underlying paper suggests a number was scraped out, perhaps transforming an entry for 14s. into 01s. Given more discoloration farther up on the page, the 5 in the entry for 0500li. has also possibly been altered from an unknown numeral. While there is no evidence that this particular document was altered maliciously, the mutability of Arabic numerals was and remained their greatest vulnerability.

The men and women of early modern England considered the weaknesses of the available symbolic system for numbers when trying to create a permanent record, and they compensated for those weaknesses by employing strategies to minimize the risk of alterations and increase the trustworthiness of their numbers. Of all the systems, Arabic numerals were the least trustworthy; the form of the numerals themselves, combined with the place-based value system, meant that the system did not lend itself well to inalterably recording numbers. Even early adopters of the Arabic numeral system often retained Roman numerals for their most sensitive numerical data—monetary costs. For more important and extensive monetary transactions, the Exchequer chose to continue using tally sticks instead of switching to a completely written system that might be less bulky and easier to store; the trustworthiness of the tally stick form still trumped the possible advantages of alternatives. Given the particular weakness of the Arabic numeral system, it should not be a surprise that many early modern English men and women were not particularly eager to exchange Roman for Arabic numerals.

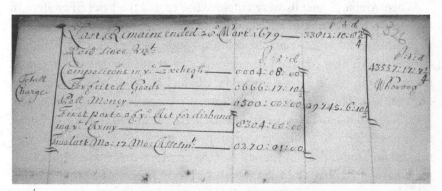

Figure 2.2 Arabic numeral account. SP46/157, f. 326r. Used by permission of The National Archives of the UK (TNA).

However, Arabic numerals were not only useful for permanently recording information; they could also be employed as a system for calculation.

Manipulability and Calculation

Unlike systems optimized for recording information, most early modern symbolic systems used for calculation allowed a certain amount of manipulation and transformation of its symbols. Performative systems worked well for calculation, and anyone who needed to verify the calculations could listen or watch as calculations were literally performed before them. Similarly, most object-based systems were well suited to calculation, including non-split tally sticks and the system of counters and counting boards. These objects were designed to be manipulated and provided a visual representation of quantities that could be transformed into other quantities through physical movement. By contrast, the only written symbols used for calculation in the early modern period were Arabic numerals.[30] As with other systems, people had to memorize a series of arithmetical facts to calculate with them. But unlike object-based systems, Arabic numerals could not be physically manipulated to short-cut this memorization process. People who were less handy with mental arithmetic had to consult an external aid such as a multiplication table, which began to be printed in the 1610s and 1620s.[31] The material characteristics of Arabic numerals worked against their use for calculation, even as the mathematical construction of the system supported and enabled it.

Spoken number words supported mental addition and subtraction, particularly through the practice of counting on, in which two numbers are added or subtracted by speaking or thinking of one number and counting upward or downward until the answer is reached; e.g., adding 5 + 3 by counting "five, six, seven, eight." Humans naturally transition from mental and verbal arithmetic to using fingers and physical objects as mnemonic aids in support of their more difficult calculations. Fingers, in particular, were always "at hand" to serve as symbolic placeholders, but beans, bones, knots, spoons, twigs, and other objects could play a similar supporting role.[32] Indeed, one of the early uses of tally sticks was actually as a calculating aid, rather than a notched and split stick intended to create a permanent record. In its original form, the tally stick was simply a stick with marks on its length, used by shopkeepers to keep track of quantities during mental calculations.[33] After split tally sticks became widespread in England, the simple, non-split tally stick remained a useful calculating tool. In one particularly famous manifestation, Napier's bones—or calculating rods—were essentially tally sticks marked with numbers in a specific (logarithm-based) pattern that enabled the user to reduce the complicated operations of multiplication and division to simpler addition and subtraction.

However, during the sixteenth and early seventeenth centuries, counters and the counting board were by far the most prominent symbolic system for calculations. Counters were concrete, easily manipulable objects, which made addition and subtraction particularly intuitive.[34] Someone other than the main calculator could also visually follow or physically assist in the movement of counters, making it possible for a small group of people to watch, comment on, or participate in the calculations as they occurred. This could be especially helpful in matters of credit and debit, as several parties needed to agree and trust that calculations had been done correctly.

Significantly, counters on a counting board also possessed the two most highly touted mathematical features of the Arabic numeral system: place-value notation and—albeit implicitly—the zero. According to a traditional strand of hagiography in the history of mathematics, Arabic numerals became the dominant symbolic system worldwide because their use of place-value notation, and an explicit expression of zero made them inherently superior to all other systems.[35] For example, historian of mathematics Georges Ifrah believed:

> In conjunction with the place-value principle, discovery of the zero marks
> the decisive stage in a process of development without which we cannot
> imagine the progress of modern mathematics, science, and technology.
> The zero freed human intelligence from the counting board that had held
> it prisoner for thousands of years, eliminated all ambiguity in the written
> expression of numbers, revolutionized the art of reckoning, and made it
> accessible to everyone.[36]

According to this and similar narratives, Arabic numerals and their explicit zero were necessary for all mathematical, scientific, and technological advances made since around the twelfth through fourteenth centuries. However, such claims begin to fall apart after even a cursory examination. For example, the shift to a written system might have "revolutionized" arithmetic, but it certainly couldn't have made it "accessible to everyone" at a time when the majority of the population was illiterate.

While idolizing the explicit zero is particularly common, scholars such as Paul Lockhart have begun to push back against the hyperbolic praise and seriously consider the affordances of abacus-style symbolic systems:

> for some reason, people seem to get bent out of shape about "the invention of zero" as some kind of landmark event, not only in the history of arithmetic but in the development of civilization itself. I say *poppycock*. I would further go on to say *balderdash*. The zero allows us to disperse with the abacus frame, that's all. It's a good idea, I grant. But it's not zero that is the breakthrough concept, it's the idea of a symbolic place-value system— frame lines or no.[37]

Like Arabic numerals, counters on the counting board were a place-value symbolic system: each counter took its value from its place on the lines and spaces of the board. Unlike Arabic numerals, with their fixed 10 symbols, counters actually had the flexibility to function not just for base-10 arithmetic, but base-12, base-20, or any other base the user could imagine. All that was necessary was for the user to declare the meaning of the lines and spaces; the counters and the board remained the same.

Counters used on a counting board also had an implicit zero, namely, any line or space with no counters placed upon it. For example, two counters placed on the ones line and hundreds line indicated the quantities of one hundred, zero tens, and a single one. This in no way required a person using a counting board to have an explicit concept of zero as a natural number because the zero was indicated by simply leaving a place empty and ignoring it, rather than by a specific action or symbol. The empty place on the counting board was functionally equivalent to the 0 in Arabic numerals. While the importance of the conceptual difference between zero as absence or nothingness and zero as a natural number can and should be debated, it's worth noting that young children in modern America view 0 as the former but "six-year-olds are actually better at solving abstract mathematical operations with zero such as $a + 0 = a$" than abstract problems using other natural numbers.[38] Early modern men and women using counting board arithmetic are therefore likely to have had a similar ability to understand and work with such mathematical abstractions, despite not having an explicit concept of zero as a natural number.

Counters and the counting board did have two major disadvantages as a calculating system. First and foremost was speed. Calculations could be slow, particularly compared to similar abacus systems—such as the Japanese soroban—or written arithmetic with Arabic numerals.[39] Someone well versed in Arabic-numeral arithmetic could calculate the solution to a problem more rapidly than someone using counters, even if the mental arithmetic necessary to support Arabic numeral calculations made them more liable to err when working with large numbers.[40] Schoolmaster John Palsgrave estimated the speed differential between the systems at six to one.[41] The primary factor affecting the speed of counter calculations was the length of time required to lay out the counters on a board, as well as the time required to move the counters back and forth during the calculations. However, for an experienced counting-board user or someone less familiar with Arabic numeral arithmetic, the length of time it took to perform each type of calculation was probably closer. Furthermore, speed took second place to accuracy, especially in the realm of financial calculations where trust in calculations was essential.

Second, each step of a calculation with counters completely and irrevocably erased the previous step in the same calculation. While it is possible to perform a calculation by iteratively adding counters to create intermediate numbers, without moving any of the counters on the board, this is space and time intensive. In practice, calculations were usually sped up and the process simplified by manipulating—and erasing the previous position of—counters already on the board. As a result, there was no way to discover, after the fact, where an error occurred in a multi-step calculation. Nor was there a way to recover the previous step if the calculator noticed an error in the middle of a calculation. In both instances, the calculator had no choice but to start over again at the beginning. It is here that the practical difference between permanent recording and calculating becomes most apparent—the very act of calculation literally erased the numerical information displayed by the counters in the previous step.

Of all the symbolic systems used in early modern England, only Arabic numerals had the potential to unite the two distinct functions of permanent recording and calculation. Like counting boards, Arabic numerals combined place-value notation with the concept of zero. There were, however, three key implementation differences between the systems. First, and most famously, the Arabic numeral zero was an explicit symbol rather than inferred through emptiness. Second, Arabic numerals had nine other written symbols, each with separate meaning, as opposed to the singular symbol of a physical counter. Mathematically, there is little difference between three counters on the tens line of a counting board equaling $10 + 10 + 10 = 30$ and the symbol 3 in the tens place of a numeral equaling $3*10 = 30$. However, abandoning the additive component of the counting board's place value system was a significant shift that appears to have made Arabic numeral numeration less intuitive to early modern calculators than counting board numeration. In *The Ground of Artes*, Recorde spent fifteen pages on the concept of "Nvmeration" with Arabic numerals, and all but one of these pages were concerned with explaining the concept of place. Recorde even included a table to illustrate this discussion to ensure that his readers had a firm grasp of the mathematical foundation of the Arabic numeral symbolic system.

Third, the written characteristics of Arabic numerals created a record of both the answer and most of the intermediate steps that led to that answer. However, if a person embraced the affordances of the system and performed calculations with the same material objects as the permanently recorded final answer, the Arabic numeral system permanently consumed significant resources—whether wood, parchment, paper, or something else—to support the act of calculation. This was especially problematic before the invention of paper mills, when paper was relatively scarce and expensive, and it has often been cited as one of the factors "delaying" the widespread adoption of Arabic numerals.[42] In the medieval Islamic world, people avoided this problem by temporarily recording the steps of the calculation on dust boards. In early modern England it was more common for people to turn to wax tablets that could be smoothed out again after the calculations were complete.[43] They thus erased their calculations, much as a counting board user did and maintained the functional distinction between the materials they used to calculate and the materials they used to permanently record their results.

This continuing distinction complicates any attempt to analyze the adoption of Arabic numeral arithmetic in early modern England. However, it is still possible to find evidence of its use—or conversely, the use of counting boards—in accounts and inventories through a careful examination of variations in the general Arabic numeral adoption patterns and through serendipitous marginalia. Initially, Arabic numerals turned up in dates—generally the year, followed by the day. Next, they appeared in quantities and prices safely sandwiched in the lines of text that form the left-hand column of accounts. Only after Arabic numerals were adopted in all other contexts were they finally used for the free-floating prices in accounts' and inventories' right-hand valuation column, along with the *summa totalis* or "sum total" of that column. This last number is particularly significant, as the sum total is the closest intersection of recording with calculation, being the sum of the numbers recorded in the column above. It is in the sum total where the majority of pattern variations occur.

For example, the accounts of the Roberts family of Sussex contains a folio from 1586 that was used as a scratch sheet to calculate a sum using Arabic numerals, indicating the scribe's familiarity with Arabic numeral calculations.[44] At the same time, the scribe used Roman numerals to record almost all the numerical information in his account book, with the exception of the year of the incarnation and—tellingly—the sum total of the account columns. The scribe was thus familiar with Arabic numerals for calculation, while retaining Roman numerals for almost all aspects of recording. This anomalous use of Arabic numerals for sum totals was not an isolated example; the churchwardens of Ashwell, Knebworth, and Prescot parishes also used Arabic numerals to record the sum total of Roman numeral columns. John Isham, a London merchant and member of the Mercers' Company, and his clerks similarly used Roman numerals to denote values in their right-hand columns and then employed Arabic numerals for the sum total.[45] Peter Wardley and Pauline White's collaborative study of 2,422 probate inventories shows the same trend in a different type of financial document; in four out of the five areas where they had complete data, their probate inventory scribes began recording sum totals in Arabic numerals several years before they adopted Arabic numerals for recording prices in the valuation columns. In the remaining area—which consisted of three parishes in Kent—the use of Roman numerals with Arabic numeral sums was preceded, three years earlier, by a single inventory that used Arabic numerals with a Roman numeral sum.[46]

This suggests that, despite the general use of Roman numerals for permanently recording most numbers, these account-keepers were using Arabic numeral arithmetic. If account-keepers used scratch paper for their calculations, then the otherwise anomalous use of Arabic numerals for a sum total indicates that the sum was calculated with Arabic numerals and not transcoded back into Roman numerals afterward. While using Arabic numerals for the sum total theoretically left the number open to alteration, the item costs could always be recalculated to check the accuracy of a disputed sum. This suggests that scribes generally adopted Arabic numerals for calculation prior to trusting them in the final, most financially critical aspects of creating a permanent record. Only after adopting Arabic numerals as a system for calculation did scribes choose to adopt a single system for all aspects of recording as well as calculation.

By a similar logic, when costs were recorded in Arabic numerals and sum totals were recorded in Roman numerals, counting boards were probably used to obtain the sum. Although the sum theoretically could have been written in Arabic numerals, the lines and spaces of the counting board coincided closely with Roman numerals. It would therefore have been easier and more secure to record the final results in Roman, rather than Arabic, numerals. This pattern suggests the use of a counting board, but it is not possible to tell whether or not this was the result of a deliberate choice on the part of a calculator who distrusted Arabic numerals.

It is important, however, not to overstate the connection between people's choice of systems for recording versus calculation. Even people who recorded everything in Roman numerals may have used Arabic-numeral arithmetic; Wardley and White reported Roman numeral accounts with Arabic numeral calculations in the margins occurring as early as 1475 in the City of London.[47] In cases where a writer chose to use scratch paper, instead of writing in the margins, there would be no evidence in the written record of the use of Arabic numerals. Even knowing that a writer had once used Arabic numeral arithmetic would not necessarily be evidence of future use; just as people switched back and forth between Arabic and Roman numerals, they could also use a combination of counting board and Arabic numeral arithmetic. Recorde justified the inclusion of a "second dialogue" on counting boards by explaining they were useful both for people who could not read and write, as well as "for the[ym] that can do bothe, but have not at some tymes their penne or tables ready with them."[48] He clearly believed that those who learned the new Arabic numerals might find themselves in situations where they would need to know how to use the older system of counting boards, switching back and forth between the two systems as circumstances required.

Marginalia, then, is the surest indicator of the symbolic system used for calculation, particularly in the case of Arabic numeral scratchwork. But just as marginalia sometimes reveals the use of Arabic numeral arithmetic in Roman numeral accounts, it can also demonstrate the opposite. In 1585–6, Dame Anne Stanhope's accountant, William Foster, maintained an account with Arabic numeral dates and quantities while using Roman numerals for prices. At the bottom of the page, he wrote the sum of those prices in Arabic numerals, which makes it probable he performed Arabic numeral arithmetic but chose not to transcode the results back into Roman numerals. However, in the bottom left-hand corners of the page, someone also recorded the results of his arithmetic in a series of dots separated by vertical slashes.[49] When those lines are considered to separate pounds, shillings, and pence—as they do in the Arabic numeral sum above—it becomes clear that the dots represent counters laid out in the auditors' form of a counting board. On the left-hand page, there are three dots for 3*li*. After the slash, there are two widely spaced dots—the one on the left representing 10 and the one on the right representing 5—above a line of four dots, yielding 10 + 5+4 = 19*s*. After the second slash, the board transitions from base 20 to base 12 and there is a dot representing 6 above four other dots, yielding 6 + 4 = 10*d*. On the right-hand page, the dots similarly transcode to 5*li*./15*s*./8*d*. confirming the Arabic numeral sum.

The account thus shows the use of four different symbolic systems: two only for recording, one only for calculation, and one for both. Number words were confined to lines of text, while Roman numerals ensured the trustworthiness of the vital financial information in the valuation column. Dates and some quantities were written in Arabic numerals without concern for mistakes, as was the sum total, which could and was later verified through the use of other symbolic systems. Either Foster, as Dame Stanhope's accountant, used a counting board to confirm the accuracy of his Arabic numeral arithmetic, or someone else—such as Dame Stanhope herself—used a counting board to audit the accounts and left written evidence of it in the form of a few scattered dots. In Dame Stanhope's household, then, Arabic numeral calculations could be trusted, but only because they had been confirmed by a counting board.

―――――

The multiplicity of symbolic systems for numbers gave early modern English men and women the ability to pick and choose among them. People particularly focused on the material characteristics of each system when choosing which to trust for any specific task. They considered systems with more permanent symbols better suited to recording numerical information and developed further strategies to increase those symbols' resistance to alteration, while systems with manipulable symbols facilitated calculations. Recording and calculation remained functionally distinct operations as people regularly employed different systems for each, choosing whichever system they knew that was optimized for the task at hand.

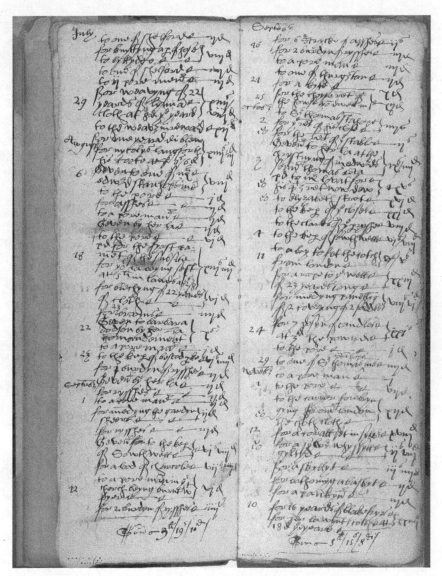

Figure 2.3 Dot notation in Arabic numeral accounts. V.b. 362, 5v-6r. Used by permission of the Folger Shakespeare Library under a Creative Commons Attribution-ShareAlike 4.0 International License.

As a consequence, Arabic numerals were not immediately seen as a significant improvement over other available symbolic systems nor were they safe from alteration and manipulation. For cases where the trustworthiness of recorded data was paramount—such as in the extensive dealings of the English Exchequer—tamper-proof tally sticks continued to be used throughout the eighteenth century. Even those who preferred to maintain written accounts tended to trust Roman, instead of Arabic, numerals for accuracy in recording the numbers in their valuation columns.

However, Arabic numerals did gain significant popularity as a calculating system during the late sixteenth and early seventeenth centuries, providing a viable written alternative to the prevailing system of counters and counting boards. For people who adopted it for calculation, the system's ability to combine calculations with the creation of a permanent record encouraged the further adoption of Arabic numerals for all aspects of recording. For those who chose to employ Arabic numerals for both recording and calculation, the advantages of the system's versatility outweighed the potential security risk.

3

"Set Them To the Cyphering Schoole": Reading, Writing, and Arithmetical Education

In late 1631, a fifteen-year-old boy named John Wallis came home from grammar school and "there found that a younger Brother of mine (in Order to a Trade) had for about 3 months, been learning (as they call'd it) to *Write and Cipher*, or *Cast account*" with Arabic numerals. Being curious, he convinced his brother to spend the next two weeks teaching him "the *Practical* part of *Common Arithmetick*," which he "shewed me by steps, in the same method that he had learned them: and I had wrought over all the *Examples* which he before had done in his book." After leaving for university, Wallis continued his education through textbooks, studying "at spare hours; as books of *Arithmetick*, or others *Mathematical* fel occasionally in my way." This education was self-directed because, as he later recalled in his autobiography, "Mathematicks (at that time, with us) were scarce looked upon as *Accademical* studies, but rather *Mechanical*; as the business of *Traders, Merchants, Seamen, Carpenters, Surveyors of Lands*, or the like . . . [and were] more cultivated in *London* than in the Universities."[1]

Wallis intended his autobiographical statements both to castigate the grammar schools and universities for the lack of mathematics in their humanist curricula as well as to extol his own virtues as a self-directed learner.[2] In doing so, he inadvertently opened a window to the educational practices of a wider, less elite segment of the English population—those merchants, craftsmen, and other account-keepers learning outside of England's Latin-based schools. Far from being extraordinary, Wallis's arithmetical education reflects a typical seventeenth-century learning experience: commercial interests inspired the introduction of Arabic-numeral arithmetic to the Wallis household through the medium of vernacular, printed books, which the Wallis brothers both read out of and wrote back into under the

By the Numbers. Jessica Marie Otis, Oxford University Press. © Oxford University Press 2024.
DOI: 10.1093/oso/9780197608777.003.0004

guidance of a more knowledgeable instructor. Significantly, for both the Wallis brothers, learning Arabic-numeral arithmetic was a fundamentally literate experience.

Until the sixteenth century, literacy and numeracy were independent skills in England, where men and women could use a variety of symbolic systems to record numerical information but primarily employed material objects to perform arithmetic. However, European commercial expansion incentivized literate merchants, tradesmen, and account-keepers to adopt Arabic numeral–based accounting practices, beginning with Italian abacists and merchants in the thirteenth and fourteenth centuries.[3] English men and women's adoption of Arabic numerals in the late sixteenth and seventeenth centuries was similarly motivated by commercial pressures and enabled by rising literacy rates and the growth of the English book trade—in particular, the introduction of vernacular arithmetic textbooks.[4] Textbooks proved to be well suited to teaching Arabic numerals' written form of arithmetic and enabled teachers to reach large audiences at more affordable prices than in-person instruction. Over the course of the seventeenth century, English educators shifted their curriculum to increasingly link numeracy with literacy and didactic literature.

Early Modern Literacy and Numeracy

Prior to the wholesale adoption of Arabic numerals, most English men and women relied on counting boards to perform arithmetical calculations. The act of calculation was predicated on the manipulation of physical objects, and it was conceptually and practically distinct from the act of recording. People who had no need to record the results of their calculations permanently, or who chose to use object-based symbolic systems such as tally sticks, could thus be simultaneously numerate and illiterate. Even people who used written symbolic systems to record numerical information were still reliant on the counting board for their calculations. While it was possible for a single individual to minimize his or her contact with counting boards by consulting pre-made arithmetical tables and "ready reckoner" books, someone still had to perform the original calculations.[5] Thus, at the beginning of the early modern period in England, there was no inherent connection between literacy and numeracy, which instead relied on one's facility with material objects.

However, the ability to write—whether with pen and paper, dust boards, wax tablets, carving, or even needle and thread—was a prerequisite skill to using Arabic numerals.[6] The timing of Arabic-numeral adoption in England suggests that this prerequisite created a prohibitively high barrier during the thirteenth through the fifteenth centuries.[7] The sixteenth and seventeenth centuries, by contrast, were a period of rising literacy rates throughout English society. Historian David Cressy used signature counting—a method that conflates the technically distinct abilities of reading and writing, withholding the status of "literate" from those who could read but not write—to estimate overall male literacy rates of only 10% in 1500. However, this rose to 30% by 1600, the same period during which people increasingly began adopting Arabic numerals for calculation, and reached 50% by 1700, when Arbuthnot hyperbolically associated Arabic-numeral arithmetic with civilization.[8] Urban areas, where merchants and tradesmen congregated, tended to be more literate than rural areas, with London and Bristol both having closer to 65% literacy in the mid-seventeenth century.[9] But in Wales, less than 20% of the population could sign their names in the late 1640s, leading to the beginning of a concentrated popular literacy drive in 1650. In Scotland, by contrast, seventeenth-century literacy rates were comparable to those in northern England, with a positive correlation between literacy and high social or economic status.[10]

While signature counting is a generally recognized method of judging literacy rates, scholars have raised significant concerns about its use in the early modern period, when reading and writing were taught sequentially rather than concurrently. These numbers must dramatically underestimate the number of people who had the ability to read, particularly those who had the ability to read printed blackletter—the "type for the common people," which was given to convicted criminals attempting to prove their literacy in order to claim benefit of clergy—as opposed to Roman type or various styles of handwriting.[11] Furthermore, judging literacy based on the ability to sign one's full name, rather than initials or a partial name, discounts as illiterate those people who had the ability to write but did not do so fluently. The extent of the possible underestimate can be seen by looking at Cressy's calculations for female literacy, which persistently lagged 10% to 20% behind their male counterparts' rates.[12] By contrast, historian Eleanor Hubbard's investigation of female literacy, which allowed both signatures and initials to stand as proxies for literacy, suggests that between 1570 and 1640, London's native female population had a 36% literacy rate and that a further 22% of its female immigrants were also literate. Significantly, Hubbard's calculations show the same rising trajectory as Cressy's, with women born before 1560 having a 16% literacy rate, rising to 29% for those born in 1600 and later.[13] Therefore, it is important to acknowledge that any specific set of literacy-rate calculations likely underestimates those with some ability to read, even as they provide a realistic approximation for the increasing percentage of the population with the prerequisite writing skills to even consider adopting Arabic-numeral arithmetic.

Vernacular Arithmetic Textbooks

Christian Europeans' knowledge of Arabic-numeral arithmetic originated in Italy's late medieval abacus schools and manuscript instructional books, called abbaci or *libri d'abbaco*. Thirteenth- and early fourteenth-century Italian mathematicians modified the high medieval rules of the Arabic-numeral system to better suit the writing materials favored by Italy's increasingly sedentary merchants and then began to teach the system to merchants through classroom lessons, which were soon also gathered into manuscripts for later reference. Arabic-numeral arithmetic proved particularly well suited to being summarized in written media, as the calculations were intended to be written in the first place. While classroom instruction predated manuscript production, the one closely followed the other; the earliest record of an abacus school comes from 1284, while the oldest of the vernacular *libri d'abbaco* still extant was written around 1290.[14] By the fifteenth century, Italians almost universally used Arabic numerals for their commercial transactions. While Italian merchants' dealings with Arabic merchants and the increasing complexity of the late medieval Italian economy might have provided the incentive for learning the new numbers, it was the abacus schoolmasters and *libri d'abbaco* that provided urban Italians with the means to do so.[15]

As printing presses began to be adopted through Europe, these new arithmetical rules and pedagogical structures moved from manuscript to print with the Treviso Arithmetic of 1478 and Luca Pacioli's more famous *Summa de Arithmetica, Geometria, Proportioni et Proportionalita* in 1494. It was at this point that *libri d'abbaco* appear to have shifted from reference books to teaching books, primarily for those who were self-educating rather than those who were paying for classroom instruction.[16] Arithmetic textbooks also began to be published throughout the rest of Europe. The first German arithmetic textbook predated the *Summa*, being published in 1482, while the first French and Spanish arithmetic textbooks were published in 1512, and the first Portuguese arithmetic textbook was published in 1519.[17] These printed textbooks must have created a dramatic increase in the number of students that a schoolmaster-turned-author could potentially reach. Some contemporary sources claimed that Italian schoolmasters taught up to two hundred students at a time and Dutch sources cite schools of up to four hundred students. However, other sources estimate concurrent enrollments of twenty-five to forty students in Italy and fifty students in Antwerp, with one schoolmaster teaching about nine hundred students during his forty-five-year career.[18] Therefore, with even a single print run of one thousand books, the schoolmaster could effectively double the number of people he taught or, if he chose to use them in conjunction with his classroom activities, produce more textbooks than his face-to-face students could use in his lifetime. Well-written textbooks could exponentially expand a schoolmaster's audience, profiting the schoolmaster and providing the self-guided student instruction at a fraction of the classroom cost.

In England, the sixteenth and seventeenth centuries saw the increasing production of printed books as a widespread, literate method of communication. More than 125,000 titles survive from the early modern period, with annual publication rising from at least 475 titles in the first decade of the sixteenth century, to over 3,935 titles in the first decade of the seventeenth century, to 18,247 titles during the chaos of the 1640s.[19] Due to questions about the survival of texts, these figures should be seen as minimum estimates for the print production in each of these time spans, rather than an accurate accounting. Bibliographer and literary scholar D. F. McKenzie argues there was a significantly greater rate of loss of texts during the late sixteenth and early seventeenth centuries, based on his quantitative analysis of London printing presses and the labor force attached to them.[20] The rate of loss for cheap print and basic educational works, such as almanacs and ABCs, is particularly high, with hundreds of thousands of copies known to have been printed during the early modern period, though only a few books are still extant.[21]

Most of the printing industry was concentrated in London, the home of the Stationers' Company and the center of English book production. London booksellers abounded and carried large inventories of diverse titles. By the 1660s, George Thomason had collected 14,942 pamphlets and 7,216 newspapers, while Charles Tias had over 90,000 octavo and quarto books in his shop, house, and warehouse.[22] There was a considerable trade in books outside of London as well. In 1585, Roger Ward of Shrewsbury had an inventory of 2,500 books, including 546 different titles; in 1615, Michael Harte of Exeter died with a stock of over 4,500 items; and in 1644, John Awdley of Hull had 832 different titles for sale.[23] Peddlers also carried smaller stocks of printed broadsides and books with them to sell at inns throughout the kingdom.[24]

The increasing availability of printed books—and the increasing percentage of the population able to read them—enabled the early modern production of over sixty unique titles related to basic arithmetic education. Approximately half of these titles were reprinted at least once, and together they went through nearly three hundred still-extant editions in the sixteenth and seventeenth centuries. Only three different arithmetic textbooks, in twenty editions, survive from the first three quarters of the sixteenth century, but their rate of production began to increase in the 1570s and continued to grow throughout the seventeenth century. Six of these titles were bestsellers reprinted twenty or more times, including four seventeenth-century titles that continued to be reprinted into the eighteenth century.[25] Judging by reprint rates, demand for arithmetic books outstripped demand for more advanced mathematical books on geometry, astronomy, navigation, and surveying; a similar pattern held for the Antwerp book market.[26] Contemporary complaints about "learned bookes [that] can not bee vnderstoode of the common sorte" generally referred to these more advanced books, such as the geometry books of Euclid, Robert Recorde, and Thomas Digges, as well as Leonard Digges's book on surveying.[27] It is likely that these specialized skills appealed to only a small subset of arithmetic students or were more often acquired through hands-on training and observation than textbooks.[28]

The first English arithmetic textbook, Cuthbert Tunstall's *De Arte Supputandi*, was written in Latin, but it was soon followed by vernacular competitors, beginning with the anonymously authored *An Introduction for to Learn to Reckon with the Pen & with the Counters*, which went through at least nine early modern editions. Unlike Italian *libri d'abbaco*, which focused exclusively on teaching Arabic numeral arithmetic, *An Introduction for to Learn to Reckon* taught both the new Arabic numeral arithmetic— "with the Pen"—alongside the established counting board arithmetic—"with the Counters"—despite the unwieldiness of needing to use multiple large printed images to portray even simple counting-board operations.[29] This textbook was thus aimed at beginners with no experience in either method of arithmetic and assumed a potential need to learn both. A few Dutch arithmetic textbooks also show this multiplicity, and it is possible that *An Introduction for to Learn to Reckon* was a direct translation of a Continental arithmetic textbook, as many early modern English schoolbooks drew from Continental sources.[30] Historian Jean Vanes claims that the printer John Herford produced this work from French and Dutch originals; however, while the use of French and Dutch mathematical terms, money, and geography suggests that the tract did have a Continental origin, Herford only printed the 1546 edition.[31] A fragment from 1526—which is possibly but not definitively the same work—was printed by Rychard Fakes and claimed to have been "Translated out of Frenshe in to Englyshe not without grete labour." In 1539, Nycolas Bourman printed his own edition.[32] A surviving copy of the 1581 edition has marginalia noting the authorship is "ascribed to W. Awdley," but that this is most probably a mistaken reference to the printer of two earlier editions, John Awdley.[33]

These early translations were swiftly overtaken in popularity by two vernacular textbooks with definite English origins: Robert Recorde's *The Ground of Artes*, first published in 1543, and Humfrey Baker's *The Wellspring of Sciences*, of 1562. Together, these two textbooks went through about seventy editions and only went out of print in the late seventeenth century, eclipsed by a new style of arithmetic textbooks that also included sections on decimal fractions, logarithms, and algebra.[34] However, marginalia in surviving editions of *The Ground of Artes* and *The Wellspring of Sciences* show that they were often resold or passed down within families.[35] For example, one 1623 edition of Recorde's *The Ground of Artes* passed through the hands of at least a half dozen annotators, including "John Griffiths" and "Mary Griffiths: his Daaghter."[36] Thus, these textbooks continued to be used throughout the eighteenth century, particularly by female students who had less access to institution-based education and were more likely to be taught arithmetic at home, if they were taught at all.[37] Recorde's books were so popular that his name became a byword for arithmetic textbooks in the seventeenth century. As grammar schoolmaster John Brinsley explained to his readers, if any of their students wished to learn more than numeration with Arabic numerals, "you must seeke *Records* Arithmetique, or other like Authors, and set them to the Cyphering schoole."[38]

As on the Continent, English textbooks emerged in the wake of face-to-face instructional activities. Recorde and Baker had both previously worked as mathematical tutors and drew on their personal teaching experiences to create a literary equivalent to the tutor—the arithmetical textbook.[39] This debt is most obvious in Recorde's *The Ground of Artes*, which was organized in the form of a dialogue between a Master mathematician and the Scholar, his pupil. The interactions between them deliberately mimicked an oral tutoring session, blurring the line between the oral and the literate.[40] The Scholar challenged the Master's authority, made calculation errors that the Master had to correct, and asked questions about complicated ideas such as the Arabic-numeral place-value system. In the original dedication of *The Ground of Artes*—made into a preface in later editions—Recorde explicitly explained his hope that the book could replace professional, face-to-face instruction for a widespread readership, particularly

> suche as shall lacke enstructers, for whose sake I haue so playnely set forthe the examples, as no boke (that I haue sene) hath done hetherto, which thyng shall be great ease to ye rude reader. Therfore good M. Whalley, though this booke can be vnto your selfe but small ayde, yet shall it be some help unto your young chyldren, whose furtheraunce you desyre no lesse than your owne.[41]

In 1582, a new edition of the book justified itself by harkening back to Recorde's preface, claiming that "it is a Booke [that] hath done many a thousand good" but that now had accumulated errors so that "when a young beginner commeth to a confused or mistaken figure, it bringeth him into a wonderful discoragement and maze."[42] The assumed lack of an external instructor to help the young beginner at such moments of confusion was part of the justification for a new, corrected edition of the text.

After Recorde's death, *The Ground of Artes* continued to be published under the editorship of mathematicians with previous practical teaching experience, who used their work editing *The Ground of Artes* as a textual method to promote their skills to an audience of merchants, tradesmen, and other literate men and women. John Dee taught mathematics in London during the 1550s, while John Mellis and Robert Hartwell both advertised themselves as mathematical tutors. Mellis styled himself a "schoolmaster" and through the 1582 to 1610 editions, he

> giueth intelligence: That if any bee minded to haue their children or seruants instructed or taught in this noble Arte of Arithmetick, or any

briefe practise thereof. [His] method is such by long custome of teaching, that (God to friend) he will bring them (if their capacitie be any thing) to their desire therein in a short time.[43]

After they gained this understanding of arithmetic, they might also learn accounts of debtor and creditor, a subject on which he also published an introductory textbook.[44] Hartwell styled himself a "Philomathematicus" and "Practitioner in the Mathematicks" who taught students in his school "In Fleetestreete, neere the Cundite, within Hanging Sword Court," which by 1632 had moved to "Great Saint Bartholomewes in the new street."[45] He advertised his arithmetical offerings in far more detail than Mellis: whole numbers and fractions, the extraction of roots, astronomical fractions, proportions, the rules of equation with algebra, and accounting. He also offered his students a variety of advanced lessons in the practical, real-world applications of mathematics that built upon basic arithmetical skills, including geometry, trigonometry, logarithms, navigation, dialing, and the use of mathematical instruments.[46]

Like Recorde, Humfrey Baker never explicitly advertised his tutoring services in his textbook, *The Wellspring of Sciences*. However, a surviving broadside that includes a detailed list of mathematical subjects Baker was able to teach, along with an example of his ability to reconcile merchant accounts, demonstrates the close correlation between his two instructional activities. The content and lesson order for both were largely similar: they began with numeration, addition, subtraction, multiplication, division, and progression in whole numbers before covering the same ground with fractions. They then continued with commerce-based applications of these operations, including the various rules of three, the rule of gain and loss, the rules of fellowship and partnership, the rules of interest, the rule of allegation, and the rule of suppositions or false positions. Both Baker's in-person arithmetical curriculum and the literate approximation of this curriculum provided by his textbook were geared toward commercial applications.

However, toward the end of his textbook and his tutoring advertisement, Baker showed his awareness of the instructional benefits and limits of the material forms of textbooks by expanding upon his standard catalog of arithmetical skills, choosing the subjects most appropriate to each informational medium. In a book that could be carried around in a pocket for reference, he included practical information on trading geared toward a merchant who might be traveling abroad and working with other merchants from across Europe. This consisted of rules about the trade of merchandise with tare and allowances, rules relating to bartering, examples of how to exchange money from one place to another, and information about weights and measures throughout Europe. By contrast, in person he taught more advanced mathematical subjects, which required more expert assistance to learn, as well as subjects geared toward tradesmen and those requiring manipulation of material instruments.[47] This included algebra; measurement of land and solid objects; the "principles of geometry, to be applied to the ayde of all Mechanicall worke-men"; the use of the quadrant, geometrical square, cross-staff, astronomer's staff, astrolabe, and Ptolemy's ruler; and double-entry bookkeeping.[48] While it was possible to teach instruments through books using paper cut-outs, these were flimsy and harder to learn on than their more durable wooden or metal counterparts.

The arithmetic textbooks of men like Recorde and Baker found a steady market during the late sixteenth and seventeenth centuries. Ownership marks in surviving textbooks indicate that these books were generally purchased within one to two years of publication. They were usually printed in the more portable and cheaper octavo or duodecimo formats. While there was no material reason for smaller books to be cheaper than larger ones, extant prices indicate that duodecimo arithmetic textbooks were sold at lower prices than octavos, which were in turn cheaper than quartos.[49] Prices for seventeenth-century arithmetic textbooks ranged from as little as sixpence to 4s., putting them roughly in line with the cost of other educational books (threepence to 3s.6d.); the slightly higher cost likely derives from the unusual typography required to print mathematical books, which tended to be produced by a specialized subset of London printers.[50] Many new arithmetic textbooks in octavo were priced about 2s.6d. in the 1630s and 1640s, rising to 3s. in the 1650s, 3s.6d. in the 1660s, and 4s. in the 1670s and 1680s.[51] Octavo textbooks that were reprints of older texts tended to be priced slightly lower than those by new authors. Robert Clavell's 1672 survey of London books priced relative newcomer *Wingate's Arithmetick* at 4s., while *Record's Arithmetick* and *Baker's Arithmetick* were 3s. and 2s.6d., respectively.[52] Textbooks in duodecimo format and used copies of octavo textbooks could be found for around one to two shillings, although one lucky student bragged of getting an even better deal and secured a witness to prove that he had come by the book honestly: "Six pence was ye price of it. Ralph Sheldon his booke witnes to it John Peeke anno dominum 1646."[53] While even this deal was more expensive that what scholars term "cheap" print—such as 2p. annual almanacs—it was a bargain compared to the 20s. an already experienced accountant paid to have an expert teach him Arabic-numeral arithmetic in 1607.[54]

Successful titles were reprinted whenever the previous edition sold out, as Thomas Rooks—the stationer who reprinted *Hodder's Arithmetick*—proudly explained in 1667:

> in this bad time of trade of Books, in less than ten months, I sold of them 1550. There being very few of this kind yet set forth by an Teacher of this Art; and as I am informed, those which are extant, of very little use to the *Learner*, without the help of an expert *Tutor*. . . . Now I desire your candid ingenuity further to observe, that these Books of the third Edition are sold and out of print, and now I present you with a 4th Edition.[55]

First appearing in 1661, *Hodder's Arithmetick* was thus reprinted three times in six years—a rate at which it continued to be reprinted for the rest of the century and into the early 1700s, despite James Hodder's death sometime in the mid-1670s.[56] The longevity of these titles, most of which continued to be reprinted long after their original authors died, indicates that popular demand for vernacular arithmetic textbooks remained high throughout the seventeenth century.

Rooks's denigration of arithmetic textbooks that required the "help of an expert *Tutor*" may have been a marketing tactic to boost *Hodder's Arithmetick*'s sales, but it also might indicate that some students had genuine difficulty learning from textbooks. While many authors attempted "to render the Rules of those excellent Arts . . . so plain and obvious, as that they may be easily apprehended without the Assistance of a living Master," others thought of their textbooks as aids to classroom instruction, such as R. B., author of a one-off textbook whose title page declared it to have been "*Designed for the use of the Free Schoole at Thurlow in Suffolk.*"[57] Others expected their textbooks to be used in a range of situations, especially toward the end of the seventeenth century, as they became part of a wider landscape of educational opportunities. Edward Cocker claimed to have addressed a fourfold audience in the 1678 edition of his *Cocker's Arithmetick*: already educated merchants, overworked schoolmasters, self-educating children, and—apparently—mathematical frauds. His description of how he expected his textbook to be used in classrooms was both bluntly honest and an intriguing window into how even the paid student might be expected to partially self-educate. He expected his textbook to serve "most excellent Professors . . . of this noble Science . . . as a monitor to instruct your young *Tyroes*, and thereby take occasion to reserve your precious moments, which might be exhausted that way, for your important affairs"—instructing their students in elementary arithmetic apparently no longer being one of them. Given the high rate at which *Cocker's Arithmetick* was reprinted for the next 150 years, Cocker must have been correct in predicting at least some of the uses for his text, although "the pretended Numerists of this vapouring age" were probably not as receptive to his text as schoolmasters and self-educators.[58]

Analyzing Arithmetical Marginalia

While arithmetic textbooks were published—and purchased—extensively during the late sixteenth and seventeenth centuries, the extent to which these textbooks aided in the learning process cannot be determined entirely from the fact of book ownership.[59] To take one example, both John Wallis and his younger brother learned arithmetic out of some sort of book: either a printed arithmetic textbook or a handwritten book of arithmetical notes, problems, and solutions that served the same purpose.[60] In the process, Wallis "wrought over all the *Examples* which he before had done in his book"; in other words, the boys left behind textual traces of their learning process in the form of marginalia. Indeed, early modern students were taught in school to mark up their books with personal notes, a process that leaves traces of students with access to additional paper for scratchwork.[61] Marginalia enables certain inferences about how arithmetic textbooks were used.

In his survey of early modern books held by the Huntington Library at the beginning of the twenty-first century, literary scholar William Sherman concluded that over 50% of surviving sixteenth-century books had substantive marginalia. For certain subjects—such as practical guides to law, medicine, and estate management—the percentage of marginalia remained over 50% for the seventeenth century as well, although the overall marginalia rate for sixteenth- and seventeenth-century books combined was just over 20%. However, Sherman argued, the number of surviving books with marginalia is only a fraction of those books that were annotated during the sixteenth and seventeenth centuries. The more heavily used books would have been more vulnerable to decay. One arithmetic textbook held in the Senate House Library contains a handwritten series of laments—possibly inserted by nineteenth-century mathematician and historian Augustus De Morgan— over the difficulty of finding a copy of the same, with similar speculations that editions have vanished because they were used until they fell apart. Moreover, many book owners had no compunction about effacing the marks of previous users, particularly by cropping off marginalia or by bleaching the pages.[62]

In my own research, I consulted 365 copies of arithmetic textbooks that survived from the sixteenth and seventeenth centuries—a little over 30% of the 1,165 individual copies listed in the *English Short Title Catalog*.[63] Of the textbooks examined, 53% contain marginalia that clearly indicate arithmetical knowledge and engagement with text, including manicules, underlining, corrections to the text, marginal glosses, and scratchwork. Arithmetical scratchwork often duplicated or expanded on the examples in the text, but it also encompassed a wide variety of other calculations, from personal accounting to determining the current age of the textbook. This reader engagement even extended to commentary on previous annotators, such as one annoyed writer who complained about the strike-outs in the text: "The book is right, and needs not this blottinge."[64] Another 23% of the textbooks contain non-arithmetical marginalia consisting mostly of owner- ship marks—such as owners' names, years of ownership, purchase dates, and purchase prices—or handwriting practice and doodles, which suggests that many books were used simultaneously to practice arithmetical and literary skills. The remaining 24% of textbooks have no markings datable to before 1800.[65]

These findings were relatively consistent across holding library, publica- tion decade, and textbook series, and the averages do not appear to have been skewed by outliers.[66] The most significant divergences appear among the dif- ferent textbook series: the textbooks of the popular Robert Recorde, Humfrey Baker, and James Hodder have marginalia rates of 87%, 80%, and 82%, re- spectively, while the textbooks of Edmund Wingate show marginalia rates of just 53%.[67] While Recorde's surviving textbooks far outnumber those of these other authors, his longevity does not distort the results; controlling for the dif- ference in publication run length, by limiting the numbers only to those copies of Recorde's arithmetic published after 1630, when Wingate's first edition was published, gives Recorde the even higher marginalia rate of 95%.

Wingate's comparatively low marginalia rates probably result from his original focus on the use of newfangled logarithms to avoid the "tedious and obscure" methods of "*Multiplication*, and *Diuision*, which so confound, and perplex the new *Practitioner*."[68] In his first edition, Wingate raced through the first five parts of arithmetic, spending longer on numeration in whole numbers, fractions, weights, and measures than on the multiplication and division that students found so confusing, which barely rated sixteen pages between them in a book of nearly five hundred pages. In second and third editions published decades later, editor John Kersey spent considerable effort bringing the text into better alignment with other arithmetic textbooks, expanding the early parts of the book so that "Learners, as desire only so much skill in Arithmetick, as is useful in Accompts, Trade, and such like ordinary employments," could find nine chapters solely on arithmetic in whole numbers "before any entrance be made into the craggy paths of Fractions, at the sight whereof some Learners are so discouraged." He also "plainly and fully delivered the Doctrine of *Fractions*, both *Vulgar* and *Decimal*," making the book "now supplied with all things necessary to the full knowledge of *Common Arithmetick*" as well as logarithms.[69] With these changes, *Wingate's Arithmetick* rated a dozen further editions before the end of the century. It is likely that readers seeking an elementary arithmetic education found these changes an improvement, as Kersey's additions increased the marginalia rate on Wingate's arithmetic considerably—only 27% of the pre-1658 editions are annotated, compared to 71% of post-1658 editions. Looking at only these post-1658 editions thus brings his marginalia rate in line with those of authors such as Edward Cocker, at 72%, as well as Thomas Blundeville and Thomas Masterson, both at 70%.

While it would be dangerous to read too much into these numbers—the sample size for Masterson is a mere ten books—it does suggest that there is a positive correlation between marginalia rates and the quality of the textbook's instruction as perceived by students. While experienced arithmeticians might purchase texts for reference or instruction in advanced techniques like logarithms, students who were still learning basic arithmetic actively marked up their textbooks. Assuming books that did not survive have a similar or perhaps even greater percentage of marginalia, then a significant number of students must have learned arithmetic from printed textbooks. It was to these students that authors and their subsequent editors particularly needed to appeal when determining a textbook's content.

Evolving Content Standards

The importance of these textbooks to the educational process can be seen in the way that their writers modified their content in response to the perceived needs of student audiences. When non-conformist minister, mathematician, and schoolteacher Adam Martindale wrote his autobiography at the end of the seventeenth century, he described his 1642 attempt to write

> a booke of Arithmetick for whole numbers and fractions in the old method of Record, Hill, Baker, &c. (for then I knew nothing of Decimals, Logarithms or Algebra) but somewhat more contracted by with an appendix of mine owne invention touching extracting the rootes of fractions.[70]

For Martindale, it was not enough to merely describe the textbook manuscript that he had lost during the civil wars as being for whole numbers and fractions. He also had to excuse the lack of subjects that by 1692 had become an expected part of the standard arithmetical textbook—the "artificial" arithmetic that consisted of decimals, logarithms, and "symbolic" arithmetic or algebra.[71]

Both Recorde and Baker's textbooks are typical of Martindale's "old method" of arithmetical instruction, which focused on introducing Arabic-numeral calculations to an English population more used to Roman numerals and counting boards. This method always began with a presentation of the first five "parts" of arithmetic: numeration, addition, subtraction, multiplication, and division. It then usually continued with progression—the construction of number sequences through addition or multiplication—and some variant of the rule of three—a memorized set of steps that enabled users to solve ratios and that was particularly useful for the conversion of currencies or the unequal division of assets. Beyond that, arithmetic textbooks' content varied, particularly with respect to the tables, practice problems, and other content that was meant to be of use in readers' daily lives. While Roman numerals occasionally appeared in these textbooks, they usually did so in the section on numeration, where the student first learned Arabic-numeral symbols. By the seventeenth century, most textbooks omitted Roman numerals entirely and related Arabic numerals back to English number words instead. The exception to this is Recorde's textbook, which, until the 1699 edition, also always included a second section on the counting boards used to perform arithmetical operations in tandem with written Roman numerals. However, neither Baker nor any subsequent authors offered instructions on counting boards, and counting boards were not mentioned in any of the printed advertisements. For the vast majority of authors, the only system they promoted and taught was Arabic numeral arithmetic.

Although the main structure of these textbooks remained the same, authors and editors constantly updated their textbooks' contents and touted the most significant changes on their title pages in hopes of convincing potential buyers that their textbooks were the most useful. Many of these additions were printed charts and tables that would have remained useful to the commercially inclined reader even after he or she mastered the basics of arithmetic. Humfrey Baker edited several editions of *The Wellspring of Sciences* before his death, and he included a variety of "most necessary Rules and Questions" aimed at an audience of merchants and artificers.[72] Starting with the 1591 edition, he also included tables of "measures and waights of diuers places of Europe" that would also be of use to merchants who traded on the Continent.[73] John Mellis added "sundry new rules" including "a third parte of rules of practice" to his 1607 edition of *The Ground of Artes*. He similarly appealed directly to merchants and traders by adding "diuerse such necessarie rules as are incident to the trade of merchandise" and "diuerse Tables and instructions that will bring great profit and delight vnto Merchants, Gentlemen, and others."[74] Of all his tables, the ones that set forth the current value of various coins must have been the most useful to readers because he specifically highlighted these tables in the title of the 1610 edition and all subsequent ones. Robert Norton added "the art and application of Decimall arithmetic" in 1615, but this failed to appeal to readers and was dropped from the next edition.[75] However, Norton's tables of board and timber measures survived and were mentioned prominently on the title pages of subsequent editions. Norton's table for 10% interest found enough of an audience that Robert Hartwell replaced it with more extensive interest tables in 1631. Hartwell's new tables were "of Interest vpon interest, after 10 and 8 per 100" as well as "the true value of annuities to be bought or sold present, respited, or in reuersion."[76]

The second half of the seventeenth century saw a striking shift in these title page advertisements, with commercially based tables replaced by subjects that would have previously been considered advanced mathematics. At the same time, the very words used to describe Arabic-numeral arithmetic in whole numbers changed from "cyphering"—a term meant to differentiate it from counting-board arithmetic—to "vulgar arithmetic"—a phrase meant to differentiate it from more advanced forms of Arabic-numeral arithmetic such as decimal arithmetic or even algebra.[77] These content changes were significant enough for Martindale to call them a whole new method of arithmetic textbook, which included arithmetic in whole number and fractions alongside decimal fractions, logarithms, and algebra.

The same changes could be seen both in textbooks and in teaching methods more generally, indicating that textual and verbal instruction remained closely linked. Robert Norton's 1615 attempt to include decimal arithmetic into *The Ground of Artes* might have failed, but Henry Phillips, editing the 1670 edition of *The Wellspring of Sciences*, proudly announced the inclusion—once again—of the "Art of Decimal Fractions, intermixed with Common Fractions, for the better Understanding thereof."[78] This time, subsequent editions continued to include Phillip's decimal arithmetic. In 1650, Jonas Moore published the first of several editions of his arithmetic, which included "ordinary operations in numbers, whole and broken" along with decimals, the "new practice and use of the logarithmes, Nepayres bones," algebra, and the mathematics of "the art military."[79] William Leybourn published his *Arithmetick: Vulgar, Decimal, Instrumental, Algebraical* in 1659, while in 1685 Edward Cocker first published his decimal arithmetic as a separate companion volume to his textbook on vulgar arithmetic. Cocker's arithmetic textbooks are particularly significant because they emerged as the new standard for English arithmetic textbooks in the eighteenth century. In 1705, when John Cannon found Recorde too antiquated to understand, he turned to Cocker instead.[80] Later in the century, Cocker's textbooks became so prominent that they even generated a figure of speech: "correct according to Cocker."[81]

Similar changes also occurred in late seventeenth-century tutoring advertisements, indicating that downplaying commercial interests in favor of a broader spectrum of arithmetical skills was part of a global pedagogical shift on the part of specialty mathematics teachers rather than a publishing trend. While tutors had offered lessons on decimals, logarithms, and algebra before, these subjects—especially decimals—began to take on new prominence in advertisements. In 1650, John Kersey, an editor of Edmund Wingate's arithmetic textbook, advertised his ability to teach arithmetic in whole numbers as well as arithmetic in three different types of "fractions"— vulgar, decimal, and astronomical. He also would teach logarithms but rated this skill less highly, burying it at the bottom of his advertisement in a note on his ability to teach the construction and use of mathematical instruments.[82]

By 1683, Henry Mose placed decimals on par with whole numbers and fractions when he simply stated his ability to teach "arithmetick in whole numbers and fractions, vulgar and decimal, and merchants accompts."[83] Also by the 1680s, Adam Martindale promised to instruct students in all parts of arithmetic: vulgar arithmetic—being whole numbers, fractions, and balancing accounts—and artificial arithmetic—being decimals, logarithms, instruments, and algebra.[84]

These changes in the standard curriculum for arithmetic textbooks and specialty mathematical tutors suggests that by the end of the seventeenth century, fewer students were seeking an initial grounding in arithmetic with a commercial emphasis. While these skills were still taught, basic lessons were insufficient to make a book or a private mathematical school profitable and had to be supplemented with lessons on more advanced and difficult subjects. It is highly unlikely that there was a sudden decrease in the need for Arabic numeral education at a time when people were increasingly adopting Arabic numerals for calculation. Instead, these changes reflected the increasing diversity of student opportunities for learning arithmetic from both formal and informal sources. In a crowded market, textbooks and tutors could stand out by teaching advanced skills as well as basic arithmetic.

Petty-School Arithmetic

Much has been written about formal education in the early modern period, generally with a focus on the changes brought about by the Reformation, humanist learning, and growing literacy rates.[85] The mathematical arts and sciences do not fare well in these narratives, as grammar schools concentrated on teaching boys Latin and Greek, while scholars considered the universities of Cambridge and Oxford to be reactionary and antithetical to the "new sciences," including mathematics.[86] This portrayal of early modern mathematical education has not gone completely unchallenged.[87] However, the overall view remains pessimistic, emphasizing the lack of a universal mathematical education rather than noting the variety of paths to mathematical training for the children who wanted or needed it, beginning at a young age.

Early modern children usually were first exposed to formal education at around the age of five or six, in what were generally called petty schools.[88] A 1406 statute enabled children of both sexes to attend petty schools, where they were supposed to acquire at least a rudimentary knowledge of reading and writing in English.[89] These schools varied widely in form, ranging from "dame schools" run by women, often informally in their homes, to "song schools" attached to great cathedrals and intended to educate the boys of the choir.[90] Falling somewhere in between were petty schools attached to grammar schools, which were intended to prepare boys for entrance to those grammar schools. These latter petty schools could be run by ushers out of the grammar school or more informally by masters in their private homes.[91] Beginning in the 1580s, petty schools increasingly became a source of education in numeracy, arithmetic, and elementary accounting.

Examples of ideal petty-school curricula can be found in printed books on elementary education. Schoolmaster and author Francis Clement first wrote *The Petie Schole* in the 1570s; his preface is dated July 21, 1576, and the work was entered into the stationer's register on July 20, 1580.[92] The original edition was focused solely on English orthography, promising "to enable both a childe to reade perfectly within one moneth, & also the vnperfect to write English aright." However, by the time of its republication in 1587, he felt the need to add—and advertise the addition of—patterns for writing secretary and Roman hands along with instructions on how "to number by letters, and figures" and "to cast accomptes, &c."[93] In this second edition he introduced his students to both Roman and Arabic numerals—including explaining the arithmetical origins of the common proverb, "but stand (we say) like a cypher in Algorisme"—but only alluded to the possibility of calculating with the latter. He felt that it would be sufficient to teach counting-board accounting, including "the due placyng, laying downe, and tykyng vp of counters" and therefore limited his arithmetical instruction to the conversion of monetary units, as well as addition and subtraction through the use of counters on a counting board.[94] Clement was not the only author to advocate teaching arithmetic to petty-school students in the late-sixteenth and seventeenth centuries. Charles Hoole, in his 1659 *The Petty-School*, desired teachers to have "good skil in Arithmetick" so that students could be taught "to read English very well, and afterwards to write and cast accounts."[95] Like Clement, Hoole prioritized the ability to read but was unclear whether he expected students to receive serial or simultaneous instruction in writing and arithmetic. The post-1550 popularity of the Rechenmeister style of Nuremberg counter, which displayed the alphabet on one side, suggests simultaneous instruction of reading and counter-based arithmetic but provides no clues as to how written arithmetic fit into curricula.[96]

Arithmetic similarly began to appear in schoolmasters' licenses in the 1580s. Of the eleven licenses reproduced by historian David Cressy in his education sourcebook, three included arithmetic alongside reading and writing.[97] In 1583, a "literatus" named Will Bradley was licensed to "teach boys the art of writing, reading, arithmetic and suchlike at Bury St Edmunds." Four years later, Thomas Cullyer of Norwich was licensed "to teach boys and infants the abc, art of reading, writing, arithmetic and suchlike." Other advertisements placed writing and arithmetic into closer proximity, grouping them together in such a way as to imply that the skills could be taught simultaneously. One 1599 license, which survived in full, authorized a fishmonger, William Swetnam of the parish of St. Margaret Pattens in London, "to teach and instruct children in the principles of reading and introduction into the accidence, and also to write and cast accounts" within the city of London.[98] At Maidenhead, Berkshire, the local chaplain "demanded but 3d a week for every scholar that learned English only, and for such as learned to write and read or to cypher or learn grammar 4d weekly."[99] There were also countless unlicensed and informal teachers, most of whom are lost to the historical record, but brief surviving mentions of their curricula often included arithmetic alongside reading and writing.[100] While it is not possible to determine how often—or how well—any of these schoolmasters taught arithmetic, the subject was becoming part of the constellation of possibilities for their students.

In the seventeenth century, a growing number of schools were founded explicitly to teach poor children literacy and arithmetic together, with the expectation that these lessons would prepare them for future apprenticeships or other honest careers. A thorough command of reading and writing in English was a prerequisite for many trades, and a "youth brought up at school will be taken Apprentice with less mony then one illiterate."[101] In an early example from 1586, George Whately of Stratford-upon-Avon set up a school in Henley-in-Arden where thirty children between the ages of eight and thirteen could learn reading, writing, and arithmetic.[102] In 1624, Sir William Borlase founded a petty school at Marlow to teach twenty-four poor children to read, write, and cast accounts; this course of instruction was expected to take approximately two years, after which the boys would have acquired the skills prerequisite to being bound as apprentices.[103] Similar schools to teach poor children to read, write, and cast accounts were founded in Beccles, Suffolk, in 1631; in Cheshunt in 1642; in Greenwich in 1643; and in Westhallam, Derbyshire, in 1662.[104] In 1694, Simon, Lord Digby, bequeathed £4 per annum to teach boys reading, writing, and accounting to prepare them for a range of future careers such as bailiffs, gentlemen's servants, or honest tradesmen.[105] The Great Yarmouth Children's Hospital, in 1696, also aimed to prepare children for apprenticeship and rewarded the schoolmaster "for teaching every child, viz., twenty shillings when it can read well in the Bible, twenty shillings more when it can write well, twenty shillings when it can cypher well to the rule of three inclusive, and twenty shillings when each girl can sew plain work well."[106]

While arithmetic gradually became an enduring component of children's early education, it is important not to overestimate the quality or universality of early modern petty-school instruction. Edmund Coote—author of *The English Schoolmaster*, which was reprinted forty-eight times over the course of the seventeenth century—focused on teaching reading and writing, with special attention to orthography for those who would go on to grammar schools. Although he felt obliged to include instruction on "the first part of Arithmetick, to know or write any number," he refused to spend more than a page on it, "my Book growing greater than I purposed."[107] Like Coote, many schoolmasters must not have taught anything beyond the most basic introduction to numeration by Roman and Arabic numerals, of the kind that could also be had from an ABC book.[108] Even those who taught addition or subtraction did not necessarily have to be skilled arithmeticians. As late as 1701, John White—the master of Mr. Chilcot's English-Free-School in Tiverton with "near Forty Years Practice in Teaching"—expounded the benefits of rote learning in his *The Country-Man's Conductor*:

> As to the Arithmetical Part (When your Children have gotten some Perfection in their English) let them learn it by heart, and if neither Teacher nor Learner understand the Use of the Rules, yet when they come to learn Arithmetick in earnest, it will be a great help to them and ease to their Master.[109]

Understanding the rules of arithmetic took second place to memorization, and White argued that even the teacher did not need to understand what he taught. White further recommended that children be taught arithmetic "before or as soon as they are put to writing."[110] This is a logical progression, given the reliance of Arabic numeral arithmetic on writing skills. However, a significant number of children, particularly in rural villages, likely dropped out of petty school after learning to read but before learning to write. Thus even those students who had access to arithmetical instruction might not have been able to take advantage of the opportunity.[111]

Grammar Schools, Tutors, and Apprenticeships

After children—theoretically—learned reading, writing, and at least basic enumeration at a petty school, boys had several different options for continuing their educations, all of which allowed for further arithmetic instruction as needed. Boys could attend grammar schools, obtain apprenticeships, or seek out other, specialized instruction such as public lectures and mathematical schools. These options were not mutually exclusive; many boys first attended grammar schools or mathematical schools and afterward were bound apprentices or gained admission to universities. Other boys traveled even more complicated educational paths. Sixteen-year-old Robert Ellison, a student at the prestigious grammar school of Eton, was supposed to begin an apprenticeship but was told that he "cannot come from thence [Eton] into a merchants' compting house without being some months at school in London to learn to write and also accounts."[112] He thus required a combination of a formal grammar school, a specialty writing and arithmetic school, and an indentured apprenticeship to prepare for his future career in trade.

As the case of Robert Ellison implies, arithmetic was not a substantial component of the continuing education of boys who attended grammar schools. The humanist curriculum of grammar schools focused on Latin, Greek, and reading the classics, none of which required a great knowledge of arithmetic, much less more complicated mathematics. Schoolmaster John Brinsley was probably only exaggerating slightly when he complained that innumeracy was "a verie ordinarie defect" and that he had seen "Schollers, almost readie to go to the Vniuersitie, who yet can hardly tell you the number of Pages, Sections, Chapters, or other diuisions in their bookes" nor "helpe themselues by the Indices, or Tables of such books."[113] While students were expected to be fluent readers, their later introduction to writing and arithmetic meant those were often less practiced skills.

Many grammar schools sought to remedy the defects in their students' petty-school educations by arranging optional extra lessons on holidays and half days. The grammar school at Rotherham offered writing and accounting lessons as early as the fifteenth century, while Bristol grammar-school students were released early on Thursdays and Saturdays for lessons with the local scrivener.[114] Statutes written by the trustees of the Blackburn Grammar School in 1597—and confirmed again in 1600—made provision for associated "petties" to be instructed in arithmetic, and schoolmasters could force grammar school students to take remedial, petty-level lessons at any time in which they were not actively engaged in their primary curriculum:

> Uppon dayes and tymes excepted from teachinge, the Scollars may be caused by the Schoole Master and the Usher to larne to write, cipher, cast accounts, singe or such licke, and allso upon holidayes, and other convenient tymes.[115]

This section of the statutes was probably enforced in practice, as the trustees also assert an unusual commitment to mathematical education in their statutes: "The principles of Arithmeticke, Geometrie, and Cosmographie with some introduction into the sphere, are proffitable."[116] A set of 1629 statutes written by Samuel Harsnet, future archbishop of York, for the Chigwell school even required one of its schoolmasters to be proficient in both writing and arithmetic in addition to Latin:

> I ordain that the second schoolmaster, touching his years and conversation, be in all points endowed and qualified as is above expressed touching the Latin schoolmaster; that he write fair secretary and Roman hands; that he be skilful in cyphering and casting of accounts and teach his scholars the same faculty.[117]

The trustees thus wanted a schoolmaster who could actively teach his students writing alongside several different mathematical skills, including the use of Arabic numerals for calculations, counters for "casting" accounts, and the bookkeeping skills necessary to record their accounts. But most grammar schools probably relied on outside tutors—ideally those who could teach both writing and arithmetic—to teach their remedial students.[118]

The common need for tutoring in writing and arithmetic increased the inherent pedagogical link between them, leading to the rise of writing-cum-arithmetic tutors who advertised their skills in both capacities. From 1582 to 1610, John Mellis, the editor of an arithmetic textbook, ran a school "within the Mayes-gate in short Southwarke nigh Battle bridge" where "children or seruants" could be taught arithmetic, accounting, algebra, and "any manner of hand vsuall within this Realme of England."[119] The Restoration, in particular, saw a significant expansion in the number of these tutors who formed their own private schools in and around London. During the 1660s and 1670s, James Hodder taught both writing and arithmetic in a house "next dore to the

Sunne in Tokenhouse Yard, Lothbury, City of London"—aside from a 1666–71 interlude in Bromley by Bow—and his school was continued by Henry Mose, "late servant and successor to" Hodder, through 1720.[120] Similarly, Edward Cocker taught writing and arithmetic from 1657 to 1676, holding classes in St. Paul's churchyard, Northampton, and then in Southwark. John Hawkins took over the Southwark school after his death, styling himself a "writing master," until his own death in 1692.[121] Hawkins's conflation of writing and arithmetic continued in his advertising for a 1680 edition of Cocker's arithmetic book, where he noted that it had been commended by "many eminent mathematician and writing-masters in and near London," implying that the opinion of a writing master should have similar value to that of a mathematician when it came to teaching arithmetic.[122] While there were fewer of these writing-cum-arithmetic tutors outside London, Cocker's Northampton school was not the only one. In 1677, Peter Perkins "taught Writing and Arithmetick, with any or all parts of the *Mathematicks* at easie Rates" near the grammar school at Guildford in Surrey.[123] Nor were all of these tutors and students male. A woman named Elizabeth Beane—who also conflated writing and arithmetic, being referred to variously as "mistress in the Art of Writing" and tutor in "The Art of Writing and Arithmetick"—taught female students in the 1680s.[124]

As with arithmetic lessons at the petty-school level, the availability of extracurricular arithmetic lessons from grammar schools and outside tutors did not mean that all students took advantage of those lessons. Local clerks, scriveners, and mathematical teachers charged fees to cover the costs of these lessons, and parents would have been most likely to pay for such lessons—particularly those focused on learning to cast accounts—in the case of sons who would eventually be bound as apprentices. However, the majority of early modern boys did not continue formal schooling by attending universities but instead left school to pursue vocational education as apprentices in trade or agricultural settings.[125] Historian Patrick Wallis estimates that, by the late seventeenth century, over 9% of England's teenaged males were serving apprenticeships in London alone, where two thirds of adult males had been apprentices in their youth.[126] It was to these sorts of apprenticeships that the poor recipients of charity education aspired, children of tradesmen flocked, and younger sons of the gentry defaulted in order to make a living.

Some form of arithmetic must have been necessary for any tradesman who expected to be paid by his customers and even more so for boys who pursued careers in carpentry, surveying, and navigation. However, mathematical instruction was not commonly specified in apprentice indentures. The relative

silence on the subject of arithmetic in indentures was probably due to an assumption that accounting would necessarily be included in any apprentice's
instruction; at least a rudimentary ability to calculate was necessary to
trade. The seventeenth-century Southampton apprenticeship registers only
record one instance in which an indenture explicitly included arithmetic. In
January 1630/1, the orphaned Giles New of Southampton was apprenticed to
a clothier who promised to instruct him "in the trade of clothier and to write
and cipher."[127] In most indentures, it must have been understood that clauses
such as "all other trades of sciences as the said [master] shall use" included
the keeping of accounts.[128] Many apprentices began their apprenticeships in
their masters' counting-houses, observing counting-house clerks perform
calculations. This was called "learning the lines," and it would enable the apprentice to eventually calculate accounts on his own. Merchants, vintners,
drapers, and haberdashers were especially likely to follow this practice.[129]

The Southampton apprenticeship register is similarly silent on the subject of other advanced mathematics that would be necessary for the practice
of specific trades. For example, no explicit mention is made of mathematical training for John Jolliffe, who was learning to be a seaman in January
of 1654/5. Given that his father was a weaver, he probably did not have extensive training in the use of mathematical instruments. Thus, additional
mathematical education would have been a necessary part of his master's
"instruct[ing] him in ye art of navigacon &c."[130] In January of 1648/9, another boy, David Jenvy, was to be instructed "in the art of merchandizing
beyond the seas. . . . Master to permitt ye apprentice to trade and trafficque
for himselfe with a stocke of 50 li when he goes to sea, which is to be in ye
two last yeares."[131] To trade overseas successfully, Jenvy needed to have
training in advanced arithmetical subjects such as the rules for commuting
and exchanging money. However, this training was understood to be part of
his general education, and there was no need to list it separately in his indenture. Whether he already had this skill before beginning his apprenticeship
or whether he obtained it by shadowing his master, reading an arithmetic
textbook, or attending lessons with a local tutor is unknown, but all were
possible routes to acquiring the arithmetic he needed for his future career.

———

In the early modern period, the previously separate skills of writing and
arithmetic were linked in English mathematical and pedagogical practices.
This connection began with the commercial impetus to adopt Arabic

numerals, which made at least some proficiency with writing a prerequisite skill to performing arithmetic. Unlike object-based methods of arithmetic, Arabic-numeral arithmetic lent itself well to both face-to-face and text-based instruction. The sixteenth century's rising literacy rates and the creation of a new genre of vernacular arithmetic textbooks made instruction in Arabic-numeral arithmetic available to any literate person—male or female—with a few shillings to spare for the purchase of a new or used book. Early textbooks sold steadily, if slowly, until a surge of interest in the 1570s and 1580s led to the publication of a host of new textbooks and a drive to incorporate basic Arabic-numeral arithmetic into the petty-school curriculum.

Over the next fifty years, account books and probate inventories show the overwhelming adoption of Arabic numerals for calculation among the literate part of the English population. The successful introduction of Arabic-numeral arithmetic to England is also reflected in textbooks and tutors' changing content standards in the mid-seventeenth century, as they began to cede more introductory lessons in Arabic numerals to a variety of formal and informal educational alternatives in order to focus on more advanced topics like decimals, logarithms, and algebra. Seventeenth-century petty schools and charity schools increasingly incorporated arithmetic into their curricula while writing-cum-arithmetic tutors—many of whom were themselves textbook authors—reinforced the pedagogical connection between writing and the pen-and-paper arithmetic of Arabic numerals. While some English men and women continued to use counting boards throughout the eighteenth century, Arabic numerals had become the dominant, literate symbolic systems for mathematical calculations.

4

"According To Our Computation Here": Quantifying Time

On the "viij⁰ die Februar' 1609, anno 7⁰ et 43 Regis Jacobi &c.," the indenture of a young man named John Newell was recorded in the Southampton apprenticeship register. Newell was to be bound an apprentice "for the Terme of seven yeres from the feast of St. Thomas Thappostle last past."[1] In other words, Newell was bound to serve his master for the seven years following December 21, 1609, of the old style, Julian calendar, which was also December 31, 1609, of the new style, Gregorian calendar. This information was written into the register a little over a month later, on February 8—or 18, new style—in the English legal year AD 1609, the Scottish legal year AD 1610, the seventh year of the reign of King James I and VI of England, and the forty-third year of his reign in Scotland.

This multiplicity of chronological systems begins to illustrate both the complexity of time during the early modern period and the way people used numbers to temporally locate themselves.[2] The clerk who enrolled John Newell into the apprenticeship register used several calendars in his efforts to pinpoint two moments that were crucial for Newell's hopes of a future career and social status in Southampton as precisely as possible. All but one of these calendars were built on an explicitly numerical foundation. Even the liturgical calendar of Christian holy days was implicitly quantitative, with holy days either permanently fixed to the numerical days of other calendars or varying according to mathematical formulas.[3] The Feast of St. Thomas the Apostle was a date, in and of itself, but it was also the twenty-first day of December.[4]

Time was highly quantified at the beginning of the sixteenth century. Numbers were an explicit part of most calendars, but advanced mathematical knowledge was required to construct calendars and calculate time based on astronomical observations. However, the way ordinary people used numbers to locate themselves in time changed over the course of the early modern period. During the late sixteenth century, religiously motivated

By the Numbers. Jessica Marie Otis, Oxford University Press. © Oxford University Press 2024.
DOI: 10.1093/oso/9780197608777.003.0005

and contested calendar reforms led people to adopt the notation of mathematical fractions to accommodate competing calendar systems. In the seventeenth century, the proliferation of almanacs brought the multiplicity of
early modern calendars directly into the household and encouraged people
to place themselves at the center of their own personal calendars. Along with
the increasing incidence of global travel, almanacs also raised popular awareness of the relativity of time and the relationship of time with space. By the
eighteenth century, calendars and time were generally understood to be first
and foremost numerical constructs, not in and of themselves sacred, but
rather important only insofar as they helped people temporally locate themselves in order to conduct their mortal and immortal lives.

Consensus and Variation in Sixteenth-Century Europe

At its most basic, time is a measurement of duration; it answers the question
"how long?"[5] The duration of an event is understood to be the length of time
between its start and end, whether that length is measured quantitatively
with numbers or qualitatively by comparison to some other, non-numerical
duration. Both "three quarters of an hour of length" and "as long as you can
say an Avemary" are measures of duration, though only one is explicitly
quantitative.[6] Time is commonly used to order events by measuring the duration between each event and a fixed reference event. These events do not
need to have already occurred, so time can be used both to record past and
predict future events.[7] Calendars are a formalized version of this; they establish an official reference event, or group of events, that are subsequently used
to chronologically order all other events. In other words, calendars assign all
events to a specific temporal location: a date.[8]

During the early modern period, temporal locations were defined using
three different conceptions of time—linear, episodic, and cyclical. Linear
time fixes a single important event as the center of its chronology and measures duration forward and backward from that event along an imagined
time line. Episodic time is similar to linear time but relies on multiple events,
which are usually all of the same type, and reckons duration by each event
only until the next event occurs. That next event then becomes the new
chronological reference point, and the process repeats itself. Finally, cyclical
time consists of an event or events that repeat in a regular pattern without a
clear beginning or end.

At the beginning of the sixteenth century, a temporal consensus existed throughout Europe that used all three types of time in conjunction with one another. Christian kingdoms, city-states, and other political entities all agreed on what constituted the basic chronological units of the calendar: the 12-month year, the 28- to 31-day month, the 7-day week, and the 24-hour day.[9] A linear calendar enumerated the years and built on a cyclical foundation of episodic months and cyclical days. The months repeated in a fixed cycle; the days of a month were numbered sequentially, but that sequence restarted with each new month; and the days themselves were built on a continuous cycle of hours with no beginning or end. While these hours could be further subdivided into minutes and seconds, in practice only specializations such as astronomy required this level of temporal precision.[10]

This shared calendar generally followed rules established in the Roman Empire by Julius Caesar's calendar reforms of 46–5 BC.[11] The so-called Julian calendar was based on the solar year, calculated as the length of time required for the sun to travel from one equinox to the same equinox the next time it occurred.[12] While the Julian calendar also incorporated chronological subunits, the solar year does not divide evenly into any naturally defined subunit. Instead, it consists of approximately 365.24 solar days and 12.37 lunar months. To accommodate the partial day in the solar year, the Julian calendar employed a "leap year" or "bissextile" by adding an extra day in the month of February every four years.[13] It did not, however, make any attempt to incorporate lunar months; instead, days were grouped into weeks and months of arbitrary length, which were derived by calculation from days and years rather than from observed astronomical phenomena. Ignoring lunar cycles immensely simplified calendar calculations, but it also meant that important Christian feasts based on lunar cycles did not exist in fixed relationship with the rest of the calendar and were thus known as the "movable feasts."

Within this general consensus, there was still room for variation. Cyclical temporal units have no inherent beginning or end, making the day and the year particularly open to interpretation. Natural astrological phenomena that occur repeatedly over the course of the day—such as sunrise, noon, sunset, and midnight—were all obvious candidates and variously used to mark the division between one day and the next, but they were not the only choices. In parts of Italy, for example, the daily cycle of hours restarted half an hour after sunset, a time that approximates the end of local civil twilight, when artificial illumination becomes necessary outside.[14]

The year was similarly open to interpretation, even under the Roman Empire. Julius's initial new year's day of March 1 was soon modified to January 1, to coincide with Roman election cycles.[15] While starting the year on the first day of a month—any month—was logical, it was also far from the only choice. After the fall of the western Roman empire, various successor states adapted a Roman practice of numbering years by the reign of their rulers, in so-called regnal calendars.[16] These episodic calendars had a new year's day corresponding to the monarch's accession day; numbered years from the beginning of each monarch's reign; usually ended with a partial year as monarchs rarely died on their accession day; and restarted with the accession of a new monarch.

In addition to the use of locally defined regnal calendars, various Christian political entities also came to agree on the use of a shared regnal calendar, one whose primary reference date was the incarnation of Christ, the King of Kings.[17] The years calculated from the incarnation were known as *anno domini*, which was translated into English as years of grace or years of our Lord, and today is abbreviated AD. By the early modern period, years of the incarnation were increasingly paired with a backward-counting system of years "Before Christ" or BC, in which the year AD 1 was preceded by the year 1 BC, which in turn was preceded by 2 BC, and so forth.[18] It was this shared calendar that allowed Europeans to agree that Bede standardized the *anno domini* calendar in the eighth century or that Henry VII became king of England in AD 1485.

The date of the new year, however, varied from kingdom to kingdom, as there was no consensus on when in the year Christ's reign began. Some retained the Roman January 1, but other Christian kingdoms instead adopted December 25—Christmas, the celebration of Christ's birth—or March 25—the Annunciation or Lady Day, the celebration of the beginning of Christ's incarnation on earth—as the logical start of the *anno domini* new year. Some even used the movable feast of Easter—the celebration of Christ's resurrection—as a new year's day.[19] Thus, while all of Christendom might have celebrated the fall of Muslim Granada to Christian forces on January 2, they would have disagreed on whether that event occurred in AD 1491 or 1492. These small temporal variations had to be accommodated when communicating across political and cultural borders, but the broader temporal consensus made such accommodations relatively straightforward. Unfortunately for the men and women of late sixteenth-century Europe, the rules of the Julian calendar were inherently flawed, and the temporal consensus was about to be broken.

Breaking Time and Broken Numbers

The Julian calendar had provided a solid foundation for ancient time reckoning, but cracks began to appear in that foundation over the following centuries. Julian leap years did not completely reconcile the Julian year with the actual length of the solar year, leaving a difference of approximately 11 minutes and 16.5 seconds. This difference did not create a discernible problem in the short term, but as centuries passed, it caused the Julian calendar to slip increasingly out of alignment with the natural cycle of the equinoxes. This primarily impacted movable feasts, the most important of which was Easter, the celebration of Christ's resurrection from the dead. In AD 325, the Council of Nicaea had officially fixed Easter to the Sunday following the first full moon after the vernal equinox of March 21.[20] By the sixteenth century there was a difference of ten days between the natural cycle of the equinoxes and the fourth-century Julian calendar that assumed a March 21 equinox. As a consequence, calculations done according to the established calendar formulas no longer matched those done according to astronomical reality, with each producing a different date for Easter. Even worse, in places like England that adhered to a March 25 start of the new calendar year, some years didn't have an Easter, while others celebrated it twice.[21]

Although the calendar's slippage led to repeated calls for reform over the centuries, it was no longer clear who had the authority to alter time. The reforms of the first century BC had been proclaimed by a Roman emperor, and as early as AD 453, Pope Leo had written to the eastern Roman Emperor Marcianus about the problem of the calendar drift, but Europe was no longer politically united.[22] In theory, Christian Europe shared a religion, and various popes considered the problem of calendar reform in the 1340s, 1430s, and throughout the sixteenth century.[23] However the twelfth-century schism that divided Christianity into two separate churches—the Eastern Orthodox and Roman Catholic—meant that any decision a pope made on calendar reform might not be respected by Christians who followed the patriarch of Constantinople. The Protestant Reformation further complicated questions of authority, but also led Pope Gregory XIII to call the Council of Trent in 1563.

This council undertook the reform of the Catholic breviary and missal, with the understanding that this would necessitate some sort of calendar reform to fix the problems associated with calculating the date of Easter. Several options were considered, including a fourteen-day reform to return the Julian calendar to its original relationship with the equinoxes and an eleven-day reform to return the calendar to its state at the time of Christ. However, the council eventually decided to make a ten-day reform to the time of Nicaea.[24] By choosing Nicaea as the basis of their reformed calendar, they intended to both emphasize Trent's Nicaea-like power as a general church council as well as encourage the Eastern Orthodox churches to adopt the same reform.[25] In February 1582—assuming a January 1 new year—the new calendar was published in the papal bull *Inter gravissimas*. Various kingdoms and polities followed this with civil decrees requiring the adoption of the new calendar, and ten months later, October 4 was followed by October 15 in Spain, Portugal, Poland, and parts of the Italian peninsula.[26] Intriguingly, the chronological unit of the week was not affected by this reform. October 4 was a Thursday and was followed by Friday—not Monday—October 15. The base cycle of Sunday holy days was thus unaffected by the reforms.

While the English acknowledged the flaws in the Julian calendar, they were reluctant to adopt the Gregorian calendar because to do so risked acknowledging papal authority over time and, implicitly, papal authority over religion. This was a particularly sticky proposition in the 1580s, given the 1570 papal bull excommunicating Queen Elizabeth and calling for all loyal Catholics to overthrow her. In December 1582, Elizabeth's Privy Council—possibly at her request—engaged mathematician John Dee to advise them on the matter.[27] Dee produced a treatise by February and testified on the matter in March. He suggested that Elizabeth proclaim her own calendar in her role as a British empress, much as the original Julian calendar had been proclaimed by a Roman emperor. This would both neatly avoid the question of accepting the pope's authority and promote Elizabeth's own authority as an imperial power.

Dee advocated this reform be set not to Nicaea but to the time of Christ, for "Christians should regard his birth as the 'Radix of Time'" and thus the proper moment to form the basis of the reformed calendar.[28] The English reform would be more correct than that of Pope Gregory, who surely must eventually recognize the superiority of the English calendar.[29] But upon further discussion with William Cecil, Lord Burghley, Dee agreed to remove ten days and conform with Catholic Europe, so long as this practical concession was accompanied by theoretical arguments that the entirety of Christendom ought to subsequently find one more day to remove from their new calendar. Whether ten days or eleven, these days were to be removed gradually over the course of several months to avoid social disruption. Elizabeth seems to have accepted this advice, as an unpublished proclamation dated April 28, 1583, declared her intent to publish a special calendar removing ten days between May and August 1584, so that by September the English calendar would be in "accord with other countryes next hereto adjoyninge beyond the seas."[30]

However, this modified proposal was scuttled after being sent for approval to Edmund Grindal, archbishop of Canterbury. Grindal joined with three other English bishops to object to the proposal using a combination of theological, political, and mathematical arguments. Just as the Catholic Counter-Reformation was at the heart of the Gregorian calendar reform, Protestantism was at the heart of the bishops' objections. They argued that the exact timing of Easter was a matter of historical convention rather than doctrinal necessity, as it was only a commemoration rather than a reenactment of Christ's death and rebirth. Even if it had not been, the Gregorian calendar retained a one-day error from the time of Christ, which meant its dates continued to be inaccurate, just like those of the current calendar. Given the choice between two calendars that were inaccurate in different ways, it was the duty of the Protestant churches to distance themselves from the Roman calendar.[31]

The bishops also rejected the idea that a monarch had the political authority to reform the calendar and insisted another ecumenical council like Nicaea would be required to make any necessary changes. On this point, Elizabeth had an obvious rebuttal in the Royal Supremacy, which included control over the ecclesiastical calendar that her father had used vigorously during the 1530s. More successfully, the bishops pointed out that calendar reform would also require a revision of the 1559 Book of Common Prayer. This could all be dealt with through a parliamentary act, so Elizabeth deferred the question to the next parliament in 1585. "An Act, giving Her Majesty authority to alter and make a calendar, according to the calendar used in other countries," was read several times but lost when the parliament was dissolved less than a month after its introduction, and with it went the last real push for calendar reform in the 1580s.[32]

Although religion and politics were sufficient to stop the calendar reform, the bishops intriguingly put forth a third argument for refusing to change the calendar: the mathematical consequences for the general population. Before submitting their opinion, they sought the advice of "some godly learned in the mathematicalls" as well as merchants who "have continual resort by cause of contracts and traffic for trade of merchandises" to the Continent. Having to accommodate different calendars would be inconvenient, but "diverse marchants of best experience inhabiting within the citee of London do think and offer to prove, that they may use their traffic as well without that alteration as with it."[33] Subtracting ten days from the current date was definitely within the capabilities of merchants who were already used to making the complicated mathematical calculations necessary to exchange foreign currencies.

If merchants did not need the calendar to be changed to support international commerce, the bishops argued, then the mathematical drawbacks of adopting the new calendar would outweigh any possible benefits:

> the alternation will ease but a few, viz.—such as have traffick with foreyn nations; but to the rest of the realm it will be troublesome. For the old rules of the compound manual of the Golden number, of the epact and cycle of the sonne, &c. whereby generally the people of this realme doe find out the course of the year, the change of the moon, and consequently the tides and the Dominical letter, &c. (which hitherto have served them) will be wholly out of use, and hardly shall they learn new, which peradventure will also be more uncertain.[34]

Thus, for the majority of the population, the calendar adhered to by the rest of Europe was irrelevant. However, any attempt to alter the English calendar would result in widespread confusion, as the old rules for calculating lunar dates would be made invalid. Without these rules, farmers could no longer correctly calculate the seasons, nor mariners and fishermen the tides. Every movable feast, fair, and law term would also have to be recalculated according to a new set of rules. While the population would eventually adjust to and memorize new rules for calculation, their grasp of these rules would remain uncertain and more prone to mistakes than their calculations by the current rules. In short, the bishops argued that people were capable of performing calculations with memorized formulas, but lacked enough understanding of the mathematics underpinning those formulas to easily adapt them to a new calendar.[35]

By the time the bishops raised their objections to the planned reform, some English men and women were already working, living, and communicating across calendrical "borders." Letters extant in the Elizabethan state papers show a minimum of fuss and confusion at the changes in late 1582. Some simply continued using the Julian calendar or adopted the Gregorian calendar without comment. Others noted the implementation of the new calendar, such as when Thomas Stokes wrote from Bruges on December 23 about how "Yesterday by proclamation from the Court, and proclaimed here in this town, 'that yesterday' was appointed to be New Year's Day and to be the first of January: so they have lost Christmas Day here for this year.—Bruges, the 23 December 1582, '*stillo anglea*' and here they write the 2 January 1583."[36] Letters began to be dated "according to the reformation of the calendar," "according to the new calendar," or "according to our computation here" with the occasional longer note such as was written by one correspondent from Strasbourg who dated a letter "according to the old style, for we do not use the new on in these parts."[37] Within a few months, people stopped giving explanations and simply flagged dates being in the English style or date, the old style, or the new style.[38] But even that appears to have been more ink than writers wished to spend on the calendar reform, and they soon began abbreviating old and new style to O.S. and N.S. in English, or st. v. and st. n. in Latin, French, Italian, and other languages, a custom that continued into and throughout the seventeenth century.[39]

The quest for clarity and concision eventually led people to co-opt the notation of mathematical fractions, also known as "broken numbers," to conveniently indicate the date in both the Gregorian and Julian calendars. This was a relatively straightforward notation, as Humphrey Baker explained in his arithmetic textbook: "Fractions or broken numbers . . . are two noumbers with a line betwene them bothe."[40] Thus Sir Dudley Carleton could date a letter written April 10 in the old style calendar and April 20 in the new style calendar, as "ye 10/20 of Ap. 1615."[41] There was no consensus as to which date should come first, the Julian or the Gregorian. Elizabeth Stuart, exiled Queen of Bohemia and daughter of King James VI/I, used both interchangeably, writing letters on "3/13 of Januarie," "22/12 of April," and "31/21 of May."[42] Fraction notation for the day could even be used in conjunction with fraction notation for the year when spanning a new year, such as a letter she received on "19/29 March 1638/9."[43]

The most complicated part of translating between the Gregorian and Julian calendars occurred when the calendars disagreed on the month, as well as the day. In these situations, some people had no trouble subtracting across the months, giving dates such as "28 Jullet / 8 Aust 1632," "1 d'Aoust / 22 de Jul. 1639," and "30 Mars / 10 April 1640."[44] However, others preferred to employ a single calendar during these transitions and simply marked their dates as being old or new style, using qualitative rather than mathematical signifiers to indicate which calendar they had employed.[45] Only when the Gregorian date had advanced past the tenth of the month did they revert to using fraction notation. They were thus extremely fluent in translating between calendars and only rarely made errors in their dates, such as one letter corrected via strikeout to read "10/~~20~~ of October N stile."[46] Navigating between the various calendars was inconvenient but well within their mathematical capabilities.

In theory, the majority of people in England remained unaffected by the Gregorian calendar reform. In practice, there was one change that had already begun to filter into England: the establishment of January 1 as de facto new year's day, as opposed to March 25 as codified in the English civil calendar.[47] Scotland even officially adopted January 1 as the beginning of the year in 1600, just prior to the union of the English and Scottish monarchies, leaving the Stuarts to reign over kingdoms with officially different calendars. The uncertainty surrounding the start of the new year grew to such a point by the middle of the seventeenth century that Lady Anne Clifford of Westmorland felt the need to explain that her birthday was "ye 30th of January (Being Fryday) in ye yeare 1590 as ye yeare begins on New Yeare's Day."[48]

Here, too, the notation of mathematical fractions could provide clarity with concision. As early as 1613, the first ledger book of High Wycombe gave the date as "22 January 1612/3" and fraction notation became more common as the seventeenth century wore on.[49] Most people used only one or two extra numerals to denote the next year, giving dates such as "Feb: 5th 1660/ 61" or "March 15 1671/2."[50] However, other people chose to write out the entire Arabic numeral year above and below the line, thus even more closely mirroring fractions: "24 ffebruary 1631/1632" and "22. March 1635/1636."[51] This notation thus acknowledged the numerical uncertainty of the calendar by conceding that a date could belong to either of two different years, depending on what day one chose as the start of the new year. While not adopted universally, it still proved extremely useful for reducing calendrical confusion, so much so that people studying the early modern period still use it today.

The late sixteenth century thus saw the shattering of the European temporal consensus. Attempts to create a universal reformation of the increasingly inaccurate Julian calendar foundered on disagreements about what type of people, or groups of people, could exercise authority over time. In

England, political and religious arguments against reform were further bolstered by evidence that the population had sufficient mathematical skill to add and subtract ten, but not to learn new versions of the complicated rules for calculating movable feasts that were not fixed to a certain day of the calendar. To navigate conflicting calendars and eliminate temporal confusion, seventeenth-century English men and women co-opted the notation of fractions—commonly referred to as "broken numbers"—to convey two different dates for the same event. The calendar had been both literally and notationally broken.

Annual Almanacs and English Times

The English adoption of January 1 as new year's day was at least partially facilitated by the new genre of the yearly almanac, which followed the astronomers' custom of beginning the year on that date.[52] Prior to the sixteenth century, most calendars were designed to work in perpetuity and required readers to either consult separate tables or use mathematical rules to calculate the movable feast days for any particular year. The first printed calendars to be prepared for specific time periods were still designed to last for decades.[53] In the late sixteenth century, however, the decreasing cost of printed books led to the rise of the yearly almanac, which achieved its standard form during the reign of Queen Elizabeth and fell under the control of the Stationers' Company by 1571.[54] Almanacs were reliable moneymakers—generating a profit of £1000–1500 annually by the Restoration and rising even higher thereafter—and the Stationers worked hard to suppress rival, illegal almanacs while minimizing costs to maximize their profits.[55] Under their monopoly, seventeenth-century almanacs grew to be second in popularity only to the Bible, selling over 400,000 copies annually in the 1660s and 1680s.[56]

At their core, almanacs consisted of a calendar and a prognostication, which helped people temporally locate themselves with respect to past, present, and predicted future events. But almanacs also contained a wide variety of historical, medical, and scientific information—including mathematical information geared toward popular consumption, such as explanations of the calendar reform, tide tables, ready reckoners for calculating interest on loans, and discussions of upcoming meteorological phenomena like eclipses.[57] Authors further used almanacs to advertise expertise as physicians, mathematical educators, instrument makers, and textbook authors, seeking to convert their readers into customers and students.[58] These almanacs appealed to all levels of society; they were relatively cheap, costing between one and six pence, and also commonly available in alehouses or other shared spaces for those who couldn't afford to buy one outright.[59] While the vast majority of almanacs must have been thrown away at the end of every year, when their calendars became obsolete, a minority were kept and placed in bound collections for later reference. This was a common enough practice that Arthur Hopton expected the readers of his 1611 almanac to be able to refer back to "the directions in my Almanack 1608."[60]

Until the 1750s, the English almanac adhered to the astronomical version of the Julian calendar, reckoned from Christ's incarnation and running from January 1 to December 31. The limited temporal scope of these almanacs meant that authors could calculate a whole year's worth of movable feasts and other astronomical phenomena, eliminating most of the reader's need to perform his or her own calculations.[61] After the Gregorian reform, many almanac writers also helped readers track the differences between the two calendars by adding dual tables of Julian and Gregorian movable feasts. This can be seen in a Dublin almanac as early as 1587 and in London almanacs by 1591.[62] By the seventeenth century, dual tables of the Julian—also known as English or Sosigenian—and Gregorian—also known as Roman or Lilian—calendars had become regular features in many almanacs, though some almanac writers provided incorrect tables in 1700 when they initially failed to realize the gap between calendars had widened to eleven days.[63] These tables generally included a list of the main movable feasts in each calendar, listed side by side for easy conversion, but also gave the necessary information—namely, the golden number, dominical letter, and epact—for readers who wished to perform their own calculations.

While English almanacs used Julian years of the incarnation as their primary calendar, their authors also sought to increase their appeal by including other popular early modern chronological systems. The Julian calendar was the dominant calendar employed by the English monarchy and elites, and it was codified as such in 1538 with the establishment of parish birth, marriage, and death registers that employed Julian dates for recording life events.[64] However, it was not the only chronological system available in early modern England. Historian Keith Wrightson's survey of early seventeenth-century deposition books and wills from rural Durham and the cities of Newcastle-upon-Tyne and Gateshead found five different types of temporal reference events: Julian dates; ecclesiastical feasts; moments in the seasonal and agricultural cycles; historical events, especially those of personal importance; and the present moment. Surprisingly, he found little evidence of people using the English regnal calendar, though it thrived in other contexts. Overall, Julian dates were used in 42.4% of wills, while ecclesiastical and seasonal references were used in 44.7% of wills.[65] Perhaps unsurprisingly, ecclesiastical and seasonal references were more common in rural areas—where agricultural cycles based on those seasons and calculated by those feasts structured daily life—than in towns, where Julian dates and comparisons to historical and personal events were more common.[66]

To ensure their broadest possible appeal and utility, it was therefore imperative for almanac calendars to not only provide readers with Julian dates but also associate those dates with seasonal events and the liturgical cycle of Christian holy days. These events and holy days were integrated into the main calendar; almanacs' monthly pages listed the major holy days next to their respective Julian dates, particularly days that commonly served a dual function as religious festivals and temporal markers for mundane civic activities. For example, the so-called Quarter Days, were ecclesiastical and seasonal events used to subdivide the year for the payment of rents. When translated into Julian months and days, the Quarter Days of Lady Day, Midsummer, Michaelmas, and Christmas became March 25, June 24, September 29, and December 25. These were, respectively, 91, 96, 87, and either 90 or 91 days apart depending on the leap year. The Quarter Days were therefore chosen for their importance as holy days, as opposed to their exact mathematical quartering of the year. The legal and university terms were also based on the cycle of holy days but were calculated from—rather than fixed to—movable feasts. Simply indicating the Julian date of these holidays on the monthly pages, as was possible to do with Quarter Days, would have left the reader responsible for calculating the term dates on his or her own. Therefore, almanacs included verbal explanations and tables in which their authors had already calculated the start and end date of that year's legal and university terms.[67]

The importance of the ecclesiastical calendar for scheduling secular events can perhaps best be seen in almanacs' continued inclusion of old saints' days that had been eliminated from the official English liturgical calendar during the Reformation. This was a situation of no little embarrassment to Protestant almanac writers, who felt the need to justify their inclusion of apparently Popish festivals. George Wharton, in his 1648 almanac, explained that some of these days are appointed by the Church to be kept holy. Others, (such as *Cuthbert, Martin, Hilary, Thomas Becket, &c.*) because the principall Marts and Faires kept here in England are by them distinguished, and the Dates of old Evidences thereby the better discovered, and not for any superstitious use, as the ignorant suppose.[68]

It was not, strictly speaking, necessary for events based on the feasts of Cuthbert (March 20), Martin (November 11), Hilary (January 13), and Thomas Becket (December 29) to refer to ecclesiastical feasts rather than Julian dates in the calendar. But it was easier for some people to remember that the Hilary Term began the Sunday before Hilary rather than the Sunday before January 13.[69] Indeed, when Great Britain eventually adopted the Gregorian calendar in 1752, it caused confusion where it altered market and fair dates set according to seasonal events and ecclesiastical feasts.[70]

In addition to incorporating references to ecclesiastical and seasonal events, almanacs included tables of information on historical events. The most important of these was the regnal table, which listed every English monarch from William the Conqueror to the present along with the *anno domini* date their reign began and the length of their reign. These enabled people to translate back and forth between years of the incarnation and English regnal years, an episodic calendar that reckoned time from the accession of each English monarch. These were commonly found in the legal system because they were officially used to enumerate Acts of Parliament. For example, the first Act of Supremacy was passed in November 1534, which was the twenty-sixth year of King Henry VIII, and the act was known as 26 Hen. 8 c. 1. Acts passed during the reign of co-monarchs Philip and Mary were referred to with respect to both their regnal years, such as 2 & 3 Ph. & M. c.9—an act against gaming passed in 1555, which was the second and third year of the reigns of Philip and Mary.

Anyone who worked in or had recourse to the legal system would have encountered the use of English regnal years, but they also appeared in accounts, letters, wills, leases, deeds, and other personal documents, particularly during the sixteenth century.[71] In his household book, Sir Edward Don used regnal years exclusively for the fourteen years from "Mensis Septembris anno H. viii ii°" to "Novembyr anno xvi^to."[72] Regnal years could appear alone, interchangeably with or side by side with years of the incarnation. Sir Henry Savile wrote several letters to William Plumpton that he consistently dated with regnal and *anno domini* years, on "xxviij of November, anno 1544, 36

H.8," "xxviij of May, anno 1545, 37 H.8," and "vth of May, anno 1546, 38 H.8."[73] The Pyrton, Oxfordshire, churchwarden accounts were more variable; they began in "the seconde yere of the reyne of owre sovereigne Lord Kynge Edward the syxt," then switched to anno domini years in 1554, returned to regnal years in "the iiij yere off the rayne off our ssoverentt lade Elysabeth," and began mixing the two in "the yeare of oure lorde god 1566 and in the 8 yeare of the rayne of quene Elzabethe [sic]."[74]

Regnal years appear to have become less common over the course of the seventeenth century.[75] However their continued importance to the government can be seen in the 1649 declaration of the House of Commons forbidding their use. The day before the execution of Charles I, the Commons passed an act to establish a "New style and form in legal proceedings, writs, etc." to erase monarchical references from all English legal proceedings. Their first concern was to remove the title of King from legal paperwork. Eliminating regnal years was an immediate, and vital, corollary. The Act declared that, within

the Kingdoms of England and Ireland, Dominion of Wales, and Town of Berwick upon Tweed, Instead of the Name, Stile, Title and *Teste* of the King, heretofore used, That from henceforth the Name, Stile, Title, and *Teste* of *Custodes libertatis Angliæ authoritate Parliamenti*, shall be used, and no other; And the date shall be the year of the Lord, and none other.[76]

Despite this call to enshrine *anno domini* years as the only English calendar, the continued appeal of a nationalist English alternative prompted the republican Henry Marten to introduce 1649 as "the first year of freedom," which reckoned successive years from 1649. This system was subsequently used for dating documents under the great seal during the Commonwealth.[77] After the Restoration, the use of regnal years was resurrected, and Charles II asserted the legitimacy and continuity of his reign by deliberately backdating his accession to the year his father died. Thus, the year 1660 was styled as "the twelfth year of our reign."[78]

While regnal years were integral to the English legal system, they were not the easiest calendars to use. A survey of the Southampton apprenticeship registers demonstrates the possible confusions and mistakes that could arise from calendar conversions, with over a dozen seventeenth-century enrollments bearing conflicting regnal years and years of the incarnation.[79] The regnal years of James I and VI and Charles I were particularly problematic as they began on March 24 and March 27, respectively, and were in close proximity to the March 25 new year's day for the year of the incarnation.[80] James also had the added complication of different new year's days for both his English and Scottish regnal years, as he ascended the two thrones on different dates.[81] March of 1624/5 would have been especially difficult for record-keepers because there were three new years' days within a five-day period: March 23 was in 1624 and 23 and 58 James I and VI; March 24 was in 1624 and 24 and 58 James I and VI; March 25 was in 1625 and 24 and 58 James I and VI; and March 27 was in 1625 and 1 Charles I.[82] Thus, it is not surprising that apprentice Thomas Bridgwater was listed as having been indentured on "xxvj die Martii 1625, anno regni regis Caroli primo," when Charles I did not actually become king until the next day, March 27.[83] While some of the errors probably arose from habit, as the writer simply forgot to increase the number for the year after new year's day, others must have resulted from confusion over the start dates of each regnal year.

Almanacs thus undertook to supply readers with regnal tables that readers could use to confidently translate half a millennium's worth of regnal dates into their corresponding year of the incarnation. Regnal tables became a standard almanac feature during the 1570s and continued to be printed throughout the seventeenth century.[84] Given the limited space available to almanac authors—generally only three sheets or forty-eight pages, with twenty-four pages reserved for the monthly calendar—the continued dedication of an entire page to the regnal table indicates that readers must have found it useful. It formed one of the standard features of the almanac, much like the Zodiac Man or the prognostication.[85] Without these expected features, potential readers would see the almanac as deficient. As almanac author Edward Pond put it, "He with contempt would straight refuse to buy This book, and 't is no Almanack contend."[86]

Intriguingly, some regnal tables also had a third column that measured the number of years "Since their reign."[87] Unlike the dates that monarchs' reigns started—which were vital for determining regnal years—or the lengths of their reigns—which computed the number of regnal years each monarch had—there is no immediately apparent reason to calculate how many years had passed since their deaths. Their death dates also corresponded to the starts of their successors' reigns and thus the table duplicated information about months and days, while performing only the simplest of calculations— subtraction—on the year of each monarch's death. By doing so, the authors inverted the usual flow of time. Rather than reckoning the date of each monarch's death in reference to some prior event, such as the incarnation of Christ, they reckoned the date of each monarch's death in reference to the present, the reader's own personal time.

Almanacs further supported readers who thought of important religious and national events in relationship to the present through their inclusion of a table of historical events, known as a chronology. Rather than being reckoned according to the year of the incarnation, events in a chronology were all temporally situated with respect to the publication date of the almanac. This was also, presumably, the year in which the almanac was being read. Therefore, the underlying question for each entry in a chronology was always "how long since." Readers were not told that a great plague decimated London in 1665; they were told that it had been six years since that plague. Similarly, a Norman duke did not conquer England in 1066, but rather it had been 605 years since the William the Conqueror's invasion.[88]

Some of the chronology events were from the world's legendary past and, while they were dated with numerical specificity, no firm consensus existed across almanacs as to exactly how long ago they had been. This could include secular events, like a notice that it had been 2,485 years since Cambridge was built—which conveniently just beat out Rome, at a mere 2,423 years old. However, they were more often religious events, such as Noah's flood or the creation of the world. The creation was especially popular and often appeared both in chronologies and on almanac title pages, such as Francis Perkin's 1671 almanac for "The year of our Lord God 1671. BEING The third after *Bissextile* or Leap-year. And from the Worlds creation 5634."[89] It is here that the discrepancies between almanacs are most apparent. John Gadbury calculated that 4,713 years had elapsed between the creation and the incarnation, while Francis Perkins thought it had only been 3,963 years. Thomas Bretnor set the difference at 3,982 years, while George Wharton and Henry Coley both argued it was instead 3,949 years.[90] Such dates might have been numerically precise, but they were by no means accurate.

However, most events fell within the span of recorded history or living memory and were generally agreed on across all almanacs. These included events of religious significance, such as the burning of Saint Paul's church or Martin Luther's opposition to the Pope, as well as events of political or technological significance. In 1679, Coley helpfully pointed out on his title page that the year could also be "Numbred From the Constitution of the *Kalendar* by *Julius Caesar* 1722," as well as from the "Reformation thereof by *Pope Gregory*. 97."[91] The internal chronologies added even more events that would specifically appeal to English readers, such as the Norman conquest, Guy Fawkes's attack on Parliament, the execution of Charles I, and the Restoration of Charles II. They also emphasized several important technological inventions, particularly the invention of the movable-type printing press that enabled the production of almanacs in the first place. By doing so, the chronology did not merely provide a brief history of England and the world. It consistently located historical events in relationship to the present and thus the reader's own, personal time.

A subset of the yearly almanacs further evolved to enable their use as a combination of account book, diary, and planner, the ultimate locus of personal time. The monthly pages of early almanacs were crammed full of dates and events, leaving as little empty space as possible on the page. Some readers would interleave the pages of the almanac calendar with extra pages that they subsequently used to record events of personal significance. These could range from noteworthy events that would become part of an annual schedule of familial celebrations, such as marriages and the birth of children, to the minutia of daily life, such as travel schedules and financial transactions.[92] Authors and publishers recognized this use. As early as 1566, one enterprising almanac maker created *A Blancke and Perpetuall Almanack* with blank pages facing each month of the calendar, designed primarily to note financial matters and "things that passeth from time to time (worth of memory to be registered)."[93] By the end of the sixteenth century and continuing through the seventeenth century, a substantial minority of almanac makers modified the format of their monthly calendar pages to include blank pages.[94]

In the almanac shown in Figure 4.1, the left-hand page is still full of useful information relating to the days of the month. Note the inclusion of festival days, the phases of the moon, and times—sometimes down to the minute— for sunrise, sunset, and various weather predictions on the left-hand page. By contrast, the right-hand page is almost entirely blank so that the reader can make diary entries next to—and thus temporally locate events in—every day of the month. This type of almanac was called a "blank" to differentiate it from the more standard "sorts."[95] Blanks enabled and encouraged readers to fill their calendar pages with events of personal significance, creating a record of their life that was firmly tied to specific moments in the passage of time.

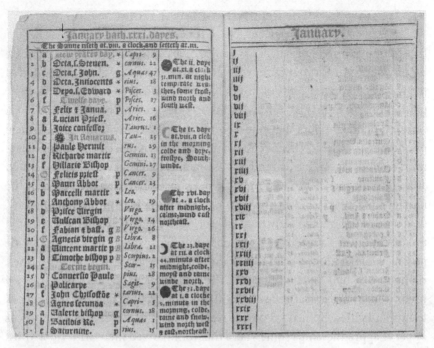

Figure 4.1 January calendar page and blank Joachim Hubrigh's annual almanac for 1565. STC 462.2 Used by permission of the Folger Shakespeare Library under a Creative Commons Attribution-ShareAlike 4.0 International License.

Late sixteenth-century England thus saw the rise of a new type of yearly almanac, an immensely popular type of cheap literature that supported a variety of methods for reckoning time. These almanacs helped people reconcile the Gregorian and Julian calendars, calculated the relationship of a year's worth of ecclesiastical feasts and seasonal events to the Julian calendar, and supported conversions between years of the incarnation and English regnal years. Almanacs further supported readers who thought in terms of present-oriented, personal time through the inclusion of chronologies and blank space intended for reader diaries. While the Stationers' monopoly theoretically gave them complete control over almanac content, both authors' comments and the development of blanks suggest that popular demand for specific features played a significant role in determining the standard form of the yearly almanac. Early modern almanacs reflected, as much as shaped, popular modes of thinking about time.

The Relativity of Local Time

In addition to providing calendars that people could use to temporally locate themselves to the year, month, week, and day, almanacs also foregrounded the smaller temporal units of hours, minutes, and even seconds with their predictions for sunrises, sunsets, weather, tides, and more. By the sixteenth century, unequal hours that could be measured directly from astronomical phenomena had largely been replaced by equal hours that could only be measured from devices such as sand glasses or mechanical clocks that operated at a constant rate.[96] The first public clocks in England were erected in the fourteenth century, and clocks were more often public, rather than private, devices throughout the early modern period.[97] In 1500 around 75%—rising to 85% by 1700—of parishes had clocks in towns where there was only one parish, while in multi-parish towns and rural parishes, a quarter of parishes, rising to 40% by 1700, had clocks.[98] The odds were good that a multi-parish town would have at least one clock, making clock time familiar throughout towns and cities. Even rural parish residents would travel to market towns or serve apprenticeships somewhere that familiarized them with clock time. Churchwarden accounts show bells were constantly being used and repaired throughout the early modern period, and at least fifty English towns had clocks—above and beyond those of parish churches—before 1640.[99] These clocks and their associated bells regularly rang out the hour so that people could count the bell strokes and determine the time of day, making the clock hour an integral part of early modern parish and city soundscapes.[100]

The hour was a useful unit of time. Early modern schools, workdays, markets, and curfews officially began and ended according to hours of the clock.[101] National, regional, and local laws all referenced clock time, including the 1563 Statute of Artificers, which set daily working hours "at or before five of the Clock in the Morning . . . untyll betwixt seven and eight of the Clocke at Night."[102] While Keith Wrightson's depositions survey found that the date was more frequently mentioned than times of day, a full 86% of Newcastle and Gateshead residents employed clock hours when they mentioned times of day.[103] Historian Mark Hailwood's examination of rural quarter sessions and church court witnesses found that the most easily recognized time markers were based on astronomical phenomena—e.g., morning, noon, or sunset—but that over 40% of the population used specific clock hours. Moreover, that percentage was relatively stable across the entire early modern period.[104] Crucially, references to clock time tended to locate events between two hours, as people recognized events as happening between the tolling of each hour's bells. Hailwood persuasively argues that clock time use thus says as much, if not more, about whether someone lived and worked within the sound of a church's bells as their understanding of clock time.[105]

But while clocks rang out the hour, ringing out the half or even quarter hour was rarer. Just ringing a twenty-four-hour day, counting the hours from one to twenty-four, required three hundred strokes a day and wore out the mechanisms faster—requiring expensive repairs more often—than ringing twelve hours twice (156 strokes), eight hours thrice (108 strokes), or six hours four times (84 strokes).[106] While twelve hours twice was most common, different early modern localities chose different hour chiming patterns and put all four patterns on their clock dials, leaving travelers to cope with the mental arithmetic of translating between them.[107] Regardless of the chosen chiming pattern, adding half hours, quarter hours, or even more precise temporal measurements would be harder on the clock than just ringing the hours. The majority of men and women in charge of these clocks appear to have considered greater temporal precision unnecessary, especially when weighed against the costs of the increased wear and tear on the bells.

Even as the majority of people in early modern England only reckoned time to the hour or large fraction thereof, almanacs predicted sunrises, sunsets, the tides, and other astronomical phenomena with exact times listed by both the hour and minute, familiarizing their readers with more granular time units.[108] However, this very precision also made it apparent to readers that such predictions did not always hold for their location, sometimes varying minutes or even hours from their appointed times. Or, as almanac writers often explained it, their almanacs served only "indifferently" for various parts of the British Isles. Instead, almanacs were calculated to serve a specific geographical location, usually "the place where the rectifier hereof was borne" or lived, and numerically described in terms of longitude and "meridian."[109] Many almanacs were printed in London and "principally referred to the Meridian of LONDON" but could be used throughout the "*Kingdoms* of *England, Scotland,* and *Ireland.*"[110] There were also regional almanacs calculated for other English cities, including York, Shrewsbury, Yarmouth, Canterbury, Gloucester, Derby, Wolverhampton, Horsham, and Reigate.[111] A few almanacs even explained to their readers how to translate the almanac into local time, providing tables and instructing the reader to "ad the houre & min. standing after the name of the place you desire according as this little subscribed table instructs thee, vnto the houre & min. of the Moones being South that day" according to the almanac's monthly pages.[112]

Global travel, particularly to the West Indies in the Americas, provided extreme demonstrations of the geographical relativity of time. Royal hydrographer and globe maker Joseph Moxon printed a treatise on geography in which he explained that the time difference between London and Jamaica "is reckoned by the access and progress of the Sun: for the Sun gradually circumvolving the Earth in 24. hours, doth by reason of the Earths rotundity enlighten but half at one and the same moment of Time." By using a globe, his readers could calculate the difference in time based on the difference in longitude and discover the answer was five hours and fifteen minutes.[113] One enterprising almanac-writer, John Seller, even wrote an almanac specifically for Jamaica, while John Gadbury included instructions within his London almanac on "how to make this *Ephemeris* serviceable to the *West-Indies*."[114] He explained the time conversions as a simple matter of subtracting six hours from the printed times, "In a thing so easie, examples are needless."[115] While there was not quite a six-hour time difference between England and Jamaica, the approximation was close enough to serve for most purposes and made the arithmetic easy. For any readers who desired more exactitude in their astronomical calculations, Gadbury encouraged them to buy his 180-year Jamaican "nativity."

Just as calendar reform forced people to confront exactly how their methods of measuring the year corresponded with the solar year, so too did almanacs force people to confront exactly how they defined and measured their days compared to natural phenomena. Unlike solstices—which occurred on the same day of the year regardless of where one lived on the globe—dawn, dusk, high tide, and low did not occur simultaneously everywhere. A ten-minute difference between dawns in Bristol and London might not be noticeable to a casual observer, especially one estimating times to the hour or half hour, but it became increasingly perceivable when pinned down to the minute in a printed almanac. To some men, such as the puritan minister Richard Baxter, the official boundaries of the day were irrelevant to how people should conduct their affairs. Indeed, it was Satan himself who caused men "to trouble themselves with scruples at what hour the day begins and ends, and the like."[116] However to others, including the puritan pamphleteer William Prynne, this was a matter of such importance to how men conducted their daily lives that he dedicated an entire treatise to the subject.[117]

The people of early modern England were already familiar with and accepted various definitions of the day. Common parlance generally referred to days as beginning at sunrise. In a 1657 almanac, Thomas Wilkinson explained, "In matters of *Law*, we the Inhabitants of *Great Britain* account the day to begin at Sun-rising, and to end at Sun-setting; wherefore, whosoever is bound to pay a sum of Money on a set or certain day, need not tender it before Sun-rise, nor after Sunset."[118] Such language implied that days and nights were two distinct chronological units that followed, one upon the other. Terms such as sennight—seven nights—and fortnight—fourteen nights—had their origin in this convention.[119] However, there is no evidence in early modern almanacs to support this as a widespread view of time. For example, the monthly calendar page of Joachim Hubrigh's annual almanac for 1565 has thirty-one days in January, numbered sequentially, with no gaps to insert thirty-one corresponding nights. It was thus generally understood that the use of the word "day" to mean the period when it was light out, and "night" to mean the period when it was dark, "doth not in this account exclude the Night before as part of this first Day, and consequently the Naturall Day consisting of Night and light."[120]

In contrast to the general population, astronomers started their day at noon, a practice that continued in nautical almanacs until the early twentieth century.[121] This had the mathematical advantage of remaining relatively constant throughout the year, as opposed to dawn and dusk, which fluctuate as the days lengthen and shorten. Still others placed the day's beginning at midnight. John Goad, a meteorologist and headmaster of the Merchant Taylors School in London, explained that there were actually two days—one natural, the other artificial. His natural day began at midnight, while his artificial day began at sunrise.[122] And in legal cases, specifically "in Indictments of Murther we reckon the day from midnight to midnight."[123] For some English divines, however, there could be no question that the day began at sunset. As it said in the Bible, "And God called the light, Day, and the darknesse he called Night: and the euening and the morning were the first day."[124] Further examination of the Bible proved that "all dayes in Scripture and divine computation do alwayes begin and end at Evening, (not morning or midnight)."[125]

Whether the day began at midnight, dawn, noon, or sunset, that still left open the question of where that astronomical event occurred. Like Gadbury, New England divine Thomas Shepard noticed the variability of local time and reassured his fellow colonists that, though the hours may differ, the day itself remained the same at all longitudes around the world. Even though "our countreymen in old *England* begin their Sabbath above 4. hours before us in new, they beginning at their evening, we at our evening, yet both may and do observe the same day." This was because "a day is not properly time but a measure of time" and those measurements necessarily differed based on one's geographical location on the globe.[126] In other words, the day need not be universally fixed with relationship to a specific point in time, consisting of the same twenty-four hours everywhere across the globe; rather, the day was a duration—one full "rotation" of the sun—that could, and naturally should, be observed differently at each longitude.

Furthermore, Shepard argued that any apparent irregularities in the nature of the day around the world were the product of faulty human observations, giving the specific example of people who gained or lost a day circumnavigating the globe. After explaining the astronomical cause of this phenomenon, he advised the traveler to simply correct his faulty observations by "cut[ing] off in his accounts that day which he hath gained by anticipating the Suns course, and so rectifie his account."[127] Significantly, Shepard justified this alteration of the traveler's account by comparing it to the Gregorian calendar reform, arguing that both actions brought a faulty calendar deriving from human observation back into alignment with natural, astronomical time.[128] None of these alterations in time should concern the religiously minded, as "the morality of the Sabbath is not built upon Astronomicall or Geometricall principles."[129] Or, as astronomer Johannes Kepler more famously said a few decades earlier, "Easter is a feast, not a planet."[130] The faithful did not need to worry about temporal alterations, be they small—such as a few minutes' time difference between various parts of the British Isles or a few hours' time difference between London and the Americas—or large—such as the day gained or lost by circumnavigating the world or as the ten days lost to calendar reform.

The regularity of clock hours made it possible for yearly almanacs to make extremely precise predictions about astronomical phenomena such as sunrises and the turning of the tides. However, this very precision enabled people to see how time varied according to geographical location. Almanacs were rectified to specific geographical areas, and while they could still be used to approximate times for the rest of the British Isles, the time difference was readily apparent in the American colonies. Religious concerns over the varying beginnings of the day, particularly the Sabbath, led seventeenth-century puritan divines to declare that the relativity of local time did not affect the universality of chronological units such as the day. Furthermore, they argued that any apparent contradictions or faults in time were due to the unreliability of human observation. Therefore, it was always acceptable to correct human timekeeping to match astronomical reality—whether it be hours, days, or the calendar. The morality of Christian holy days was not reliant on astronomy. Time, as a measure of duration, was more a matter of mathematics than religion.

———

At the beginning of the sixteenth century, English men and women used a variety of qualitative and quantitative methods to measure time, which operated within the larger framework of a Europe-wide temporal consensus based on the Julian calendar reform of the first century BC. This system was built on a series of complex mathematical calculations but could be used by anyone with the ability to number the twelve days of Christmas or to count the tolling of increasingly common parish church bells to determine the hour of the day. While not explicitly religious in construction, the calendar played a key role in the religious controversies of the sixteenth century, which led to the breaking of Europe's temporal consensus in 1582.

Late sixteenth- and seventeenth-century men and women adopted the notation of fractions or broken numbers—a concept generally taught only after arithmetic in whole numbers—to help pinpoint themselves within the broken consensus. The wildly popular genre of yearly almanacs further supported conversions between the Julian and Gregorian calendars, as well as other temporal systems defined by ecclesiastical feasts, monarchs' reigns, significant sociopolitical events, and even personal time. In conjunction with increased global travel, almanacs also made clear that time was not fixed but rather varied across the globe. The seventeenth-century calendar was still important to the practice of the Christian religion, but it was no longer the sole province of popes, bishops, and synods. Time was becoming a series of numbers that ordinary men and women might need to adjust—by minutes, hours, or more—as they went about their daily lives.

At the turn of the eighteenth century, the question of calendar reform was again raised in England and other Protestant countries, spurred by the prospect of 1700 being a leap year in the Julian calendar, but not the Gregorian. In England, the calendar reform failed due to mostly familiar arguments about the religious and political consequences. However, over the next forty years calendar reform remained a topic of public discussion—a discussion that was no longer centered around religion but rather around the practicalities of being temporally at odds with the Continent.[131] In the 1740s, those who supported calendar reform found an advocate in the Earl of Chesterfield, who drew up a calendar reform bill over the course of 1749–50—*An Act for regulating the commencement of the year, and for correcting the calendar now in use*. While the reform would be of some inconvenience, this was presented to Parliament as a mathematical inconvenience, not a religious one. The bill passed easily, with little debate, and the reform itself was uncontentious.[132]

In 1752, England's calendar was finally reunited with that of its Continental neighbors after over a century and a half of schism. The new calendar was disseminated throughout Great Britain and its colonies in almanacs, which helped people navigate through the arithmetical intricacies of the Gregorian calendar and its relationship to older calendars.[133] The compromises and jury-rigged solutions to make those calendars compatible resulted in the adoption of odd dates—neither astronomically determined events nor holy days, but rather what would have been holy days in the old calendar—to mark the new cycles of festivals, ceremonies, and terms. Perhaps most famously, the tax year remained fixed to the old new year's day to avoid impacting revenues, which is why the United Kingdom's tax year still begins on April 6 today. The "broken" time of the early modern era was eventually repaired, but the old fractures left their mark.

5

"It is Oddes of Many To One": Quantifying Chance and Risk

On August 11, 1634, "a Captain, a Lieutennant, and an Ancient . . . of the Military Company in Norwich" set out on a tour of England. During their seven weeks of adventures, they passed through Cumbria and became completely lost upon the fells. They traveled "through such wayes, as we hope wee neuer shall againe, being no other but climing, and stony, nothing but Bogs, and Myres, or the tops of those high Hills, so as we were enforc'd to keepe these narrow, loose, stony, base wayes." Eventually, they "happily lighted on a good old Man" who gave them directions to escape the fells, for otherwise "It was a hundred to one, that wee should so escape this eminent land danger, as this good old Man made it plainly, and euidently appeare to vs: well, through his help (thanks be to God) wee escaped."[1]

The travelers' analysis of the danger they had faced and the likelihood of their rescue drew upon two different ways of viewing the world. First, they expressed the risk of their deaths on the fells in terms of quantified chance. They set the "odds" of their survival, without assistance, at 100 to 1. This group thus saw numbers as an appropriate and effective way to describe and comprehend the risks they had faced. Yet, when their unlikely rescue did arrive in the form of directions from an old man, they thanked God. The encounter with an old man was not something whose odds needed to be quantified because God willed them to be rescued. They were saved due to God's providence, not chance.

None of the military trio saw anything amiss in juxtaposing the language of quantified chance and divine providence, despite their derivation from two different worldviews. A probabilistic mode of thought, which relied on quantified chance, was closely associated with gambling—or, in early modern terms, gaming.[2] This was a necessary precursor of, but not equivalent to, the mathematical theory of probability developed in the late seventeenth and eighteenth centuries.[3] Rather, a probabilistic mode of thinking acknowledged the existence of chance, which followed natural and quantifiable laws,

By the Numbers. Jessica Marie Otis, Oxford University Press. © Oxford University Press 2024.
DOI: 10.1093/oso/9780197608777.003.0006

even if the arithmetic necessary for dealing with complicated real-world situations did not yet exist. In comparison, a providential mode of thinking saw God's active, directing will in all earthly events. God's providence could not be limited by natural laws as God was free to intervene in human affairs whenever and however he desired. Despite their potential differences, these two ways of thinking coexisted not only in the minds of three military adventurers, but also in the popular culture of seventeenth-century England.

Gaming was a socially pervasive activity in early modern popular culture that provided the language and context for the development of a probabilistic concept of quantified chance. Over the course of the sixteenth century, quantified chance emerged in gaming language through the term "odds," which English men and women subsequently began to apply to situations in their everyday lives. Quantified chance came into conflict with puritan views of divine providence at the turn of the seventeenth century, leading to pamphlet wars and public debates over the nature of chance and providence. The two worldviews were reconciled through a subtype of providence that explicitly relied upon natural—and hence potentially quantifiable—laws, rather than God's active intervention in worldly affairs. Early modern men's and women's growing belief in quantified chance and its ability to predict future events can be seen in the adoption of insurance outside merchant circles and the rise of public mortality statistics. But chance was quantifiable only because God had made the world to function accordingly. It was, ultimately, by the grace of God that early modern men and women could use numbers to determine the likelihood of future events and protect themselves from chance disasters.

Gaming and Chance

During the early modern period, men, women, and children all played what are today called games of chance. These game were ubiquitous—found in alehouses and coffeehouses, in public houses and private homes, in schools and on the streets.[4] Dice games were the oldest of these forms of play, dating back to at least the Roman occupation of Britain and likely earlier.[5] These games were almost purely dependent on chance, and the wildly popular game of hazard—an ancestor of modern craps—even gave one toss of the dice the linguistically on-the-nose name of a "chance."[6] Table games like backgammon and chess, while still including an element of chance, were more skill-based games and were also adopted across the social spectrum after being imported to England circa the twelfth century.[7] By contrast, card games were relative newcomers that only began to appear in the records in the early fifteenth century but quickly became as popular as older games of chance.[8] As early as the 1460s, Edward IV's Parliament attempted to ban the importation of foreign game pieces to encourage domestic manufacture.[9] During the sixteenth and early seventeenth centuries, thousands and later tens of thousands of decks of cards were being imported annually, but this was miniscule compared to the third of a million packs produced domestically every year in the early seventeenth century or the over a million a year produced by the middle of the century.[10]

While it was not necessary to bet on the outcomes of games of chance, many did so, with stakes ranging from stones and nuts picked up off the ground to counters, cards, or coins of real monetary value.[11] A host of late medieval and early modern statutes prohibited games of chance, but by the sixteenth century, these laws were more concerned with the financial consequences of gaming than gaming per se; they included exceptions for Christmas gaming and for the wealthy to play year-round.[12] Royal support for gaming can be seen as early as the 1520s when the Groom Porter—the official in charge of providing tables, chairs, and other accoutrements for the king's lodgings—was tasked with obtaining cards and dice to entertain the royal household. By the end of the century, the Groom Porter also had the right to license game houses in London, and his primary responsibility had become supporting court gaming.[13] Queen Elizabeth held an unsuccessful fundraising lottery in 1567–9, while Charles I officially incorporated the Worshipful Company of Makers of Playing Cards in London in 1628.[14] By the early seventeenth century, "any Nobleman or Gentleman of a hundreth pound lands per annum [could] licence his seruants at his discretion" to play games of chance outside of the Christmas holidays. Even those without noble sponsorship were legally allowed to play for meat and drink.[15] There were short-lived attempts to suppress gaming under the Commonwealth, but gaming came back in full force at the Restoration; a 1664 gaming law was once again focused on limiting financial risk by setting the maximum amount of money that could be bet in any one gaming session and establishing specific fines for cheating.[16]

The financial consequences of winning or losing incentivized early modern men and women to keep a lookout for cheaters, and it facilitated an implicit understanding of probability among dice players. The repeated tossing of a six-sided die—which had six equally likely outcomes when the die had not been tampered with—lent itself particularly well to this. Philosopher and historian of science Ian Hacking's experiments with ivory dice held in the Cairo Museum of Antiquities found even the most "irregular looking ones were so well balanced as to suggest that they had been filed off at this or that corner just to make them equiprobable."[17] Medieval scholars created lists of all possible outcomes of dice throws, and early modern gamers were quick to accuse others of cheating when a die showed the same number too often or too infrequently.[18] Sir Nicholas L'Estrange recorded one seventeenth-century dice game that erupted into an argument when the two players disagreed on the numbers that had briefly been shown:

> Sir William then replied, "Thou art a perjured knave; for, give me a six-pence, and if there be a four upon the dice, I will return you a thousand pounds"; at which the other was presently abashed, for, indeed, the dice were false, and of a *high cut*, without a four.[19]

Sir William's repeated observations of the dice allowed him to realize that the dice were behaving improbably; he didn't attribute this to God's active interference in the game but rather concluded the dice had been tampered with. He was certain enough of his conclusions that he wagered an enormous sum of money on an examination of the dice—specifically, he offered the cheater 40,000 to 1 odds.

During the early modern period, even gamesters who preferred to bet on tables, cards, and other entertainments that mixed chance and skill began to use quantitative odds to understand the risks they were taking with their money. The *Oxford English Dictionary*'s earliest recorded use of the word "odds" dates only to the beginning of the sixteenth century and originally referred to quantitative differences rather than probability.[20] In war, the odds were the difference in size between two armies; in voting, the odds were the difference between those for and against; and in gaming, the odds were the difference between the amounts each of two parties staked on a bet. In the first two contexts, odds were generated through simple counting of pre-assembled groups of people. In a gaming context, however, the odds were not the amount of money gamesters brought to the table but rather how much of that money they were willing to bet on achieving their desired outcome. The numerical difference in stakes was soon conflated with a quantitative assessment of the likelihood of an event's outcome, and the term "odds" was used for both. By the turn of the seventeenth century, the term was regularly employed in its predictive rather than its differential sense, and it appeared in a variety of contexts, including political, military, philosophical, and religious texts.[21]

These new predictive odds were quantitative, in the sense of being expressed numerically, but they were not mathematical, in the sense of having been calculated according to the as-yet undeveloped rules of probability theory. They demonstrate a probabilistic way of viewing the world that accepted numbers could be used to quantify the likelihood of future events, even if no one had yet formulated mathematical rules to make those numbers accurate. This lack of concern about accurate calculations can particularly be seen in mixed quantitative and qualitative statements, e.g., "oddes of many to one,"[22] which simultaneously embraced the rhetorical power of quantitative and predictive odds while dispensing with any need to assign actual numerical values to the odds.[23] But mathematical shortcomings in practice did not undermine the belief that chance existed and could—theoretically—be described and understood quantitatively when one was predicting the likelihood of unknown or future events.

God and the Gamesters

For the Christian population of early modern England, God's providence was supposed to determine the course of all earthly events, especially those whose outcome was uncertain.[24] This providence was twofold, reflecting God's status as both omniscient and omnipotent. On the one hand, God stood outside time and thus one aspect of his providence was knowledge of the past, present, and especially future. On the other hand, God was all-powerful and used that power to direct events ranging from awe-inspiring miracles to the simple roll of a die. God may have created the world in number, weight, and measure, but the relationship of his divine providence to a quantified chance was less than certain.

Many English clergymen did their best to discredit beliefs that they interpreted as undermining the supremacy of God's providence. William Gouge, minister of St. Ann Blackfriars, London, denounced the popular belief in fortune, which led to "those things which are done by God, [being] attributed, then to *fortune*, or *chance*, or *lucke*, (for these are but severall titles which are used to set out one and the same thing.)"[25] Clergymen also sharply condemned the popular astrology and prognostications of yearly almanacs, which used the course of the stars and planets to make predictions about the future and was predicated on the idea that heavenly bodies had a direct influence on earthly events, including human actions and decisions.[26] One of the most scathing condemnations of almanacs and astrology came from theologian William Perkins, who denounced writers for their inability to actually deliver true predictions and buyers for believing human prognostications, which demonstrated their "distrust in God [and] Contempt of the prouidence of God, in not reuerently regarding it."[27] While royal injunctions in 1559 and 1568–9 did restrict almanac writers and printers from publishing potentially seditious political predictions, they did nothing to curtail the general practice of astrology, and almanacs continued to include prognostications through the eighteenth century.[28]

A belief that was less overtly problematic—but no less troublesome to some—could be found in the idea of a natural order. George Herbert, rector of Bemerton near Salisbury, complained of "the great aptnesse Countrey people have to think that all things come by a kind of naturall course." In his treatise on country parsons, Herbert lamented how difficult it was for the parson to teach his parishioners that it was not feeding and caring for their cattle that led to milk, nor sowing and tending the ground that made corn grow, but rather God's providence, which possessed manna-like properties. Only his "sustaining power . . . which he continually supplyes" kept farmers' crops growing; "without [this] supply the corne would instantly dry up."[29] God's sustaining power was continuously present, as opposed to his governing power, which God used to create disasters such as fires or earthquakes.[30] Thus it was God, not nature, that deserved the credit for a farmer's crops and livestock.

However, while English clergymen were quick to defend the supremacy of God's providence, some did allow that providence was not completely antithetical to the existence of a natural order. Like Herbert, most divided the active component of providence into two parts—a general, sustaining, or universal providence, and a special, governing, or particular providence. The former referred to God's maintenance of the world and could be interpreted to include a natural order established during the Creation. This general providence operated through secondary causes; God did not act directly but rather employed intermediary forces, including so-called laws of nature. However, God was not in any way limited by these laws. He could, and did, violate them at will and function as a primary cause of earthly events. Thus God's special providence encompassed his dealings with his chosen people and his direct actions to punish or reward, particularly miraculous interruptions of the natural order. Not all clergymen agreed on the mechanisms of providence and the division between secondary and primary causes. They also argued about the implications of God's omniscience, particularly whether or not miracles were improvised interventions to keep God's plan for humanity on track, or a pre-existing part of that plan.[31] But this distinction between God as a primary or secondary cause provided a potential theological justification for the belief in natural laws and quantified chance and was a key point of contention in several well-publicized debates on gaming, chance, and providence.

At the turn of the seventeenth century, within the context of increasing differentiation between the religious practices of puritans and prayer-book Protestants, popular gaming became a flashpoint for those with puritan leanings.[32] In 1594, clergyman James Balmford launched an all-out assault on gaming in a short treatise intended to prove "the vnlawfulness of playing at cards or tables, or any other game consisting in chance."[33] His arguments were entirely rooted in the concept of providence. Pure chance, such as the drawing of lots or throwing of dice, presupposed God's special providence; thus, dice games "tempt[ed] the Almightie by a vaine desire of manifestation of his power and speciall prouidence." He then sought to prove, by logic, that "if Dice be wholly euill, because they wholly depend vpon chance, then Tables and Cards must needes be somewhat euill, because they somewhat depend vpon chance."[34] While it was true that such games were popular among the nobility and royalty, that was no excuse for allowing such abuses of God's providence to continue, and he advocated for all games of chance to be banned.[35] A decade later, puritans blamed the 1603 outbreak of plague on God's anger at gamesters' abuse of his providence, and some went so far as to claim that God ended the plague in response to the mayor of Northampton's campaign against gaming and alehouse-haunting.[36] Throughout the reign of James I and VI, puritans took such a firm stance against certain forms of recreation—including gaming—that James rebuked them in his 1618 Declaration of Sports.[37]

But not all puritans were convinced that games of chance were inherently sinful. In 1619, a London pastor named Thomas Gataker published a sermon on chance and the theological propriety of gaming that had been well received when he preached it to his congregation.[38] He took Balmford's treatise as a starting point, but instead of focusing on games, he addressed the underlying assumption of Balmford's argument: activities that involved lots were theologically unlawful because they relied on chance, which relied in turn on

God's special providence. In deconstructing this idea, Gataker argued that most lots relied on God's general providence, as enacted through natural, secondary causes. He agreed with Balmford that it was indeed unlawful to frivolously call upon God's special providence in casting a lot. However, the lots in games of chance were not cast with any assumption of God's direct intervention.

To make this argument, Gataker first distinguished between three different categories of events—necessary, contingent but not casual, and contingent and casual.[39] A contingent event was uncertain and variable, "not directed or determined by the skill, counsell, or fore-craft" of man and "not effected and produced by knowne naturall causes," as opposed to a necessary event, which was certain, determined, and predictable.[40] Contingent events could be further divided into two types: those that were and were not dependent on chance or "casualty." Under this schema, a lot was defined as a chance event purposely applied to the deciding of some doubt.[41] However, there were other kinds of chance events, which "befall men so continually in the whole course of their liues, which yet come not the most of them within compasse of a Lot . . . [such as] lighting on some one in the street or at the market, whom they desired to speak with, while they are going about other business."[42] Thus, while lots were the canonical example, they formed only one part of a larger category of chance events.

Gataker then claimed that God's general providence directed earthly events. Thus it was also general providence that directed chance events, including lots.[43] To clinch his argument, Gataker proposed the following proof by contradiction:

> if in euery Lot there be necessarily an immediate worke and prouidence of God, then it is in the naturall power of man to make God worke immediately at his pleasure: for it is in mans power naturally to cast Lots at his pleasure. But to say that it is in mans power naturally to set God on working immediately at his pleasure, is absurd. There is not therefore an immediate worke and prouidence in euery Lot.[44]

To argue that God's special providence directed the outcome of every lot was to argue, by implication, that a man had the power to force God to be the immediate, primary cause in every roll of dice or random drawing of a lottery ticket. This assertion was patently absurd to Gataker, and thus "it may be truly said, that it was Gods will, to wit, his disposing will, that the Lot should go as it hath gone."[45]

Although most of Gataker's lots relied on God's general providence, he did differentiate between lots intended to simply decide some matter of contention and those intended to divine God's will on the matter.[46] The former he called ordinary lots, while the latter were extraordinary lots, since "extraordinary power or prouidence is required for the direction of the action to that end wherevnto it is applied."[47] It was the intention of the lot user that differentiated an ordinary from an extraordinary lot, which did rely upon God's special providence. Gataker condemned extraordinary lots as presumptuous, unlawful, and "as it well deserueth, to the dunghill."[48] However, he defended gaming as ordinary lots that were determined not by the active intervention of God but rather by a "natural power" inherent in every creature and hence a product of God's creation and general providence.

Offended by Gataker's treatise and sermons—particularly his "confut[ing] me *by name* in open pulpit"—Balmford published an extensive rebuttal that consisted of a reprint of his original treatise and a "modest reply" to Gataker's arguments.[49] While willing to adopt the distinction between ordinary and extraordinary lots, Balmford continued to insist that both were guided by God's special providence.[50] To further his argument, Balmford expounded on an analogy between oaths and lots. "As an Oath, in the nature therof, supposeth the testifying presence of God: so a Lot, in the nature thereof, supposeth the determining presence of God."[51] Thus, just as an oath sworn before a legal court, by definition, implies God's approval and verification of a witness's truth, a lot cast to resolve a controversy, by definition, implies God's approval of the lot's resolution. Given this necessary connection between God's special providence and the use of lots, Balmford argued that lots could not be used for non-religious ends, any more than prayer or swearing oaths.[52] Thus gaming of any sort—whether dice games wholly consisting of lots, or card games partially consisting of lots—should be shunned for its vain appeal to God's special providence.

Writing at incredible speed, Gataker composed his own reply to Balmford's 1623 "modest reply" in less than three weeks, dissecting Balmford's arguments point by point to demonstrate "the insufficiencie of his answers" and "the imbecillitie of his arguments."[53] Gataker's central argument remained unchanged: that lots were chance, or casual, events and as such were no different in causation from other types of chance events.

What is there in the casuall falling of the Dye, or dealing of the Cards, more than in the fall of a Coyte, or lighting of an Arrow neerer or further, or the turning of a Boule, to ensnare the Conscience? Art more ruleth the one, and Nature the other; Gods prouidence and concurrence being equall in either.[54]

Gataker explicitly linked together all chance events to argue that, if gaming was blasphemous, then shooting at game while hunting or counting the number of flies circling an unattended plate of food should be considered equally evil. Through reiterating the presence of chance in nature, Gataker concluded once again that "to impose a speciall and immediate worke of Gods prouidence, vpon the casuall euent of the Lot, more than vpon other naturall accidents and deliberate actions of men, is a temerarious and groundlesse assertion."[55]

While Balmford did not dignify Gataker's second treatise with a direct reply, it is unlikely that he was ever convinced. Two years later, in 1625, he reprinted his 1603 treatise on the infectiousness of plague in which he discussed the ways God's providence manifested during plague epidemics. While he admitted that "there be causes both naturall and diuine" for surviving the plague, he castigated those survivors who "obscure Gods prouidence by attributing thine escape to this, that the plague is not infectiue."[56] Most intriguingly, he also claimed that it was unlawful for clergymen to go among the sick for the purpose of Christian charity, using familiar arguments about God's special providence:

A wanton or vnnecessary putting of God to the manifestation of his power or speciall prouidence, is a tempting of the Almighty . . . to runne into danger of the plague without necessary cause, as they do, who resort as boldly and freely to them that are sicke of the plague, as to those that are sicke of any other disease, is wantonly and vnnecessarily to put God to manifest his power and speciall prouidence in preseruing them from the Plague: therefore to runne into danger of the plague without necessary cause, as they do, who resort as boldly, &c. is a tempting of the Almighty.[57]

It was thus God's special providence that was responsible for averting the plague, specifically, and risks in general. Consequently, in both the gaming debates and in discussions about plague, Balmford equated taking unnecessary risks with tempting God.

Despite Balmford's disengagement, Gataker did not get the final word. The debate over chance, providence, and gaming continued into the next decade as Gataker's arguments were formally debated by Cambridge students and came under attack by the extremist puritan theologian William Ames.[58] Enough criticism of gaming and other popular entertainments remained that Charles I felt the need to reissue his father's declaration on sports, causing a wave of protests from puritan clergymen. Yet at the same time, other people began to adopt Gataker's views on chance. As early as 1633, clergyman John Downe reproduced Gataker's arguments in his own defense of lots.[59] While Gataker's arguments did not immediately defeat older arguments against gaming, they did provide people who wished to reconcile chance and providence with an alternative, natural-law based framework for thinking about the nature of chance.

Clergymen's attempts to explain the nature of contingent events relied on the splitting of God's providence into a special and general providence, with the latter allowing natural laws to be the secondary causes God used to direct earthly events. Gataker used this concept of general providence to defend the theological propriety of gaming during the early seventeenth century. In doing so, he unyoked chance from the active intervention of God and argued for the existence of natural laws that governed chance in all things, from throwing dice to unexpected encounters in the marketplace. At the same time, by arguing for the interchangeability of gaming and other chance events, he provided theological justification for both quantified chance in gaming and its more general use in people's daily lives. While quantified chance was applicable to any real-world situation, it would prove especially useful for calculating risk and, consequently, for risk aversion. The popular adoption of quantified chance for risk aversion was particularly apparent in the seventeenth-century mainstreaming of insurance and growing public interest in mortality statistics.

Risk and Insurance

The insurance of trade, particularly maritime cargo, was a long-standing practice in the early modern mercantile community. The distant origin of maritime insurance lay in bottomry—borrowing money against a ship, which would be repaid only if the ship safely made port—which had been practiced by the ancient Babylonians, Greeks, and Romans.

Medieval merchant guilds provided a form of mutual assurance for their members, while non-mutual forms of maritime insurance appeared in Europe around the thirteenth century.[60] By the fourteenth century, maritime insurance policies were flourishing on the Italian peninsula. Two of the earliest surviving insurance policies, from Genoa and Palermo, date to 1343 and 1350, respectively. Both Florence and Genoa enacted fourteenth-century insurance statutes to regulate these policies and keep merchants from running afoul of the 1236 papal prohibition against charging interest on maritime loans.[61]

By the early fifteenth century, Italian merchants carried maritime insurance to Southampton, England. Records exist of a lawsuit surrounding a 1426 insurance policy issued by eighteen Italian merchants, while the Borromei bank ledgers attest to at least nine insurance policies written up in the 1430s.[62] Italian merchants primarily spearheaded and underwrote these policies, which were almost all written in Italian until the 1550s. The English insurance market relocated to London in the middle of the sixteenth century, at which point native merchants began underwriting most policies themselves; they had pushed almost all foreign merchants out of the local market by the 1580s.[63] Although insurance remained a niche concern of merchants, policies were still numerous enough for Queen Elizabeth to establish a special Chamber of Assurances, which enabled merchants to settle insurance disputes without having to resort to expensive and lengthy court suits.[64] At the turn of the seventeenth century, maritime insurance was so well established that one late Elizabethan law could declare "it hath been time out of mind an usage amongst Merchants, both of this Realm, and of foreign Nations, when they make any great Adventure (especially into remote parts)" to take out an insurance policy on their ship and goods.[65]

Maritime insurance policies were focused primarily on protecting merchants from the loss of trade goods due to storms, piracy, and other misadventures, but it was only a short conceptual step from insuring goods to insuring human lives. Slaves—as human "goods"—were commonly insured in the late medieval Mediterranean, from both maritime and non-maritime hazards, such as death in childbirth.[66] Creditors also began increasing the security of their loans by taking out policies in the name of their debtors, or by having debtors take out policies that named the creditors as beneficiaries.[67] While technically life insurance, both types of policies were more about providing insurance for someone's financial interest in another person's life—the market value of an enslaved person or the amount of a creditor's loan—than about calculating and insuring the value of the life itself. But merchants also began insuring their own lives during high-risk travel, adopting language from ship hull insurance policies and applying it to their own bodies. Such policies were time-limited and required health exams or had waiting periods of forty days before the policy kicked in to ensure that sick and dying people didn't attempt to scam the insurers.[68] But unlike ships' hulls, which had a clearly defined value, merchants did not have a market price for their own bodies. In choosing to put a quantitative value on their lives, they were effectively betting on their own lives and deaths.

On the Continent, insuring the lives of slaves, debtors, and travelers morphed into thinly—and not so thinly—veiled wagers, and life insurance became heavily associated with gaming in the sixteenth century. One could purchase a whole host of "wager insurances" on even the lives of strangers in whom the policyholder had no direct financial interest—including, much to their dismay, kings and queens. This conflation of insurance and gaming led to a transnational movement against life insurance in the 1570s and 1580s, which culminated in a series of laws that made life insurance largely illegal on the Continent.[69]

In England, life insurance survived this backlash because it remained an obscure practice limited to the mercantile community.[70] Domestic slavery was not widely practiced in late medieval and early modern England, obviating the need to insure enslaved people's lives.[71] Although the early modern scarcity of coin led to the widespread use of credit to keep the economy functioning, English debts were more often reciprocal or forgiven as acts of charity than one-way relationships that needed to be insured to guarantee repayment.[72] The oldest surviving example of a non-maritime life insurance policy was issued to a London slater in the relatively late year of 1583, while the sample six-month life insurance policy listed in William West's popular legal handbook, *Symboleography*, bore the date of May 14, 1596.[73] Furthermore, West only included sample insurance policies beginning with the second edition of his handbook, in 1603. These policies didn't even form part of the main handbook, but were tucked away at the end in an addendum of miscellaneous subjects for use in "Merchants Affairs." The section on merchant affairs was included in subsequent reprints throughout the first half of the seventeenth century, but it remained distinct from the main body of the handbook where more widely used documents, such as last wills and testaments, could be found. There was enough of a growing interest in regulating life insurance that the Chamber of Assurances lost jurisdiction over it to the common law courts by 1649, but it still remained primarily a merchant affair.[74]

The first widely adopted form of insurance in seventeenth-century England was actually a novel form of insurance that hewed closer to maritime policies. In the aftermath of the 1666 Great Fire of London—which burned down more than 13,200 houses, along with dozens of churches and civic buildings—insurers began to offer policies to protect rebuilt houses against future fires.[75] As one 1680 insurance scheme explained, since "the Original of Assurances amongst Merchants" was an agreement that "a Loss might be made Good, and divided amongst Many; which otherwayes might have fallen to a Particular Person, or some few Persons, to his, or their great Detriment, and Ruin," a logical parallel might be drawn between maritime disasters and land-based ones. For example, "The Demolishment by Fire of the City-Buildings, may be of like Detriment, and Impoverishment to those Proprietors" as the destruction of a merchant's ship and cargo. Thus "Reason and Experience . . . will direct the like Security to be admitted of, for Re-building and Repairing the Casualties of Houses by Fire, to those Proprietors and Inhabitants; both being Adventurers at a Hazard, though in a different kind and Element."[76] A history of mercantile insurance laid the foundation for the widespread, public adoption of insurance, but the disaster of the Great Fire both triggered that adoption and channeled it into a very specific form.

Seventeenth-century fire insurance schemes had two main components: an assessment of risk in quantitative terms through the size of the insurance premium and a series of actions taken to reduce the overall risk of fire for their policyholders. The calculation of premiums took into account both the value of the house and the materials with which it was built. The former was a relatively straightforward acknowledgment of the value of the goods—the house—being insured. The latter, however, was based on the underwriters' assessment of the risk of the house burning, expressed quantitatively. It was widely accepted that brick houses fared better in a fire than wooden ones. As one 1679 proposal for a joint-stock company explained, "The Casualties which may happen by *Fire* amongst the Timber-Houses, are in a greater Proportion, more Violent, and Hazardous" than brick house fires.[77] Nicholas Barbon—who established the first successful fire insurance scheme in an effort to protect the houses he had rebuilt after the Great Fire of 1666—set the terms for his policies at "*sixpence per pound rent* for brick houses, and twelvepence for timber."[78] Around the time of Barbon's scheme, the City of London created a scheme that assessed its premiums based on the value of the house, not the rent. However, the City's scheme agreed with Barbon's assumption that a timber house was twice as risky an investment as a brick house and thus should be charged twice as high a premium.[79] These numbers may not have been derived through a mathematical analysis of house fires, but both early modern insurers and the people who bought their policies agreed that the risk of building in timber could be expressed and understood in quantified terms.[80]

The second component of fire insurance schemes was the establishment of fire-fighting brigades to mitigate the overall risk faced by all their insured properties. In other words, they viewed the risk of fire as something natural that could be warded against rather than the inescapable will of God. As early as 1638—in the wake of another fire that destroyed a large portion of London Bridge—a pair of gentlemen named William Ryley and Edward Mabb presented a petition to Charles I, in which they laid out a detailed plan to prevent the "many great losses by the lamentable fires hapned in and about the Citty of London." In addition to insuring houses, the company would establish a continual fire watch, provide fire engines, and establish reserves of water to aid in fire-fighting efforts.[81] While their petition failed, they anticipated the fire brigades that would become a component of all the fire insurance companies founded in the wake of the Great Fire. Companies established their own fire brigades and "maintained them in livery with badges" to fight fires throughout the city.[82] Regardless of who sponsored a brigade, the firemen's fee—anywhere between 6*p.* and 2*s.*6*d.*, depending on the size of the fire—was paid by whatever company had insured the building they saved, or sometimes by the city if a building was uninsured.[83] These costs generally formed only a small amount of the insurance premiums. For example, the Sun Fire Office limited itself to spending approximately 3% to 5% of its premiums on fire-fighting efforts, reserving the rest for insurance claims and shareholder dividends.[84] In this way, early fire insurance both mitigated the financial effects of a house fire—the risk of which was quantitatively assessed through insurance premiums—and supported fire brigades in an effort to reduce the risk of disaster.

During the last decades of the seventeenth century, England finally experienced the kind of boom in popular insurance that the Continent had experienced more than a century earlier. A half dozen fire insurance schemes were established in London alone, and one company—the Amicable

Contributors—insured over 13,000 houses by 1708.[85] By 1694, insurance policies had grown so numerous that the Crown sought to raise revenues by imposing a stamp duty on them. Not all of these were fire insurance policies, as the popular adoption of fire insurance seems to have encouraged the adoption of other insurance policies as well. While not as common as fire insurance policies, sixty different life insurance schemes formed and issued policies between 1696 and 1721.[86]

The English insurance boom echoed the abuses found in the Continental boom. Many of these new insurance schemes were de facto wagers rather than an attempt to protect something in which the insurer had a viable financial interest. In 1702, Daniel Defoe complained that "WAGERING, as now practis'd by Polities and Contracts, is become a Branch of *Assurances*; it was before more properly a part of Gaming," and he estimated that no less than £200,000 had been wagered on the outcome of the second siege of Limerick in 1691.[87] Sieges proved a particularly popular subject for wager insurance policies; during the Nine Years' War, various insurance policies offered 30% premiums against the Netherlands city of Namur succumbing to its besiegers by the end of September in 1695. During the eighteenth century, it was possible to purchase insurance on the likelihood of births, marriages, cuckoldry, highway robbery, the fall of a besieged city, and even death by drinking gin.[88] Only in 1774, with the passage of the Gambling Act, were wager insurance policies curtailed.

As popular interest in all types of insurance increased, mathematicians worked with mortality statistics to put life insurance premiums on a firm mathematical basis for the first time. Early premiums for all types of insurance were based on educated guesses that insurers adjusted according to the season of the year and reports of piracy, location and building materials, or age and health. Both insurers and policyholders agreed that risks could be gauged quantitatively, but they lacked a method for precisely calculating those risks. In 1671, a Dutchman John de Witt became the first person to attempt to calculate life annuities based on mortality statistics. In England, mathematician Edmund Halley performed similar calculations in the 1690s, using information gleaned from Silesian bills of mortality.[89] De Witt and Halley's use of the bills was a new innovation that had the potential to set life insurance on firm mathematical ground for the first time. However public mortality statistics also had a long history of being analyzed for a much simpler type of risk analysis—tracking plague epidemics.

Plague and Bills of Mortality

Plague first entered England in 1348 as part of an epidemic that killed approximately one third of the population of Europe.[90] There were reservoirs of plague in England throughout the late medieval and early modern periods, which regularly spilled over into the human population, causing both epidemics and smaller outbreaks of plague mortality. Some of England's best-documented epidemics were in its cities—particularly London, which suffered six major epidemics in the century between 1563 and 1665–7 and lost an estimated 225,000 people to plague.[91] Though plague died out in England after 1679, the Great Plague of Marseilles triggered English fears of renewed epidemics in the early eighteenth century, and there were outbreaks in the Mediterranean and eastern Europe until the turn of the nineteenth century.[92]

Unlike various Italian city-states, whose governments enacted plague policies early on and kept lists of people who had died from the plague, England did not have an official royal plague policy until the sixteenth century.[93] In 1517, London suffered an epidemic of sweating sickness that infected the Lord Chancellor, Cardinal Wolsey, and other members of the court, followed by an outbreak of the plague. Upon recovering, Wolsey drafted England's first royal policy to combat infectious disease, which was proclaimed on January 13, 1517/8 and reissued in various forms during subsequent outbreaks.[94] These early efforts came to include reports on London mortality. But unlike their Continental predecessors, the London reports emphasized the number of the dead over the identities of the dead.[95] From their early years, they were quantitative, rather than qualitative.

While it's unclear exactly what the first reports looked like, those that survive from 1528 are full of numbers—listing the weekly number of corpses in each London parish, how many of those were plague deaths, subtotals of plague and non-plague deaths, and parishes that were clear of the plague.[96] Mortality reports were circulated to the London mayor and aldermen as well as to the monarch and Privy Council, and were issued often enough in the early decades of the century that a 1555 London ordinance could require the clerks to provide the mayor with a weekly mortality report "in like manner and custom as heretofore hath been accustomed."[97] John Stow's *Annales*, published in 1580, contains numbers related to the 1563 epidemic, and it's probable that he obtained them from manuscript sources such as the ones alluded to by the London ordinance.[98] However, such reports were never widely available, and a century later, John Graunt had to omit the 1563 epidemic from his seminal analysis of London's bills.[99]

This interest in local mortality was not confined to London. Outside of the capital city, the 1538 establishment of nationwide parish registers made it theoretically possible to begin tracking every christening, marriage, and burial in England—at least those that occurred within the official church.[100] The civic annals of Bristol reported only "great plague" and "mortality by pestilence" in 1544–5 and 1551–2, but city officials began to collect mortality statistics in the second half of the century. They entered numerical death tolls into the annals in 1565, 1575, and 1603.[101] In 1579, the city of Norwich suffered an epidemic that was worse than any other epidemic since the Black Death of 1348–9; the Mayor's Court responded by having clerks compile weekly mortality reports out of their parish registers. These reports continued almost without interruption until 1646 but only differentiated plague deaths from other kinds of death beginning in 1590.[102]

Mortality statistics continued to circulate in manuscript form until the plague epidemics of 1592 and 1603, when they were first printed and made available to the general population.[103] The 1603 outbreak actually lasted for eight years, and popular demand for the printed bills was high enough that they were published continuously thereafter.[104] During the 1603 outbreak, the Exeter civic annals also reported numbers derived from "the Church books and printed tickets," which suggests that printed bills of mortality were circulated in that city, though currently there are no known copies still in existence.[105] They were probably similar in form to the London bills, which included a parish-by-parish list of total deaths and plague deaths, along with running tallies of christenings, deaths, plague deaths, and infected parishes. London's weekly bills were later supplemented annually with a general account of the preceding year, published on the Thursday before Christmas.[106] Over the next several decades, more London-area parishes were added to the bills, including the parishes of neighboring Westminster, until the "Bills of Mortality" formed a recognized geographical unit that included all of London and its suburbs.[107]

During the seventeenth century, there was sustained popular interest in acquiring either the bills of mortality themselves or at least the quantitative information contained within them. Individual bills could be purchased for a penny apiece, or families could purchase an annual subscription for 4s., a price that appears to have remained consistent across the seventeenth century.[108] London preacher Francis Raworth included them in some of his

A generall Bill for this present year,
ending the 19 of *December* 1665. according to the Report made to the KINGS most Excellent Majesty.

By the Company of Parish Clerks of *London*, &c.

	Buried	Pla.		Buried	Pla.		Buried	Pla.		Buried	Pla.
St A'bans Woodstreet	100	121	St Clements Eastcheap	28	20	St Margaret Moses	38	25	St Michael Cornhill	104	52
St Alhallowes Barking	514	330	St Dionis Back-church	78	37	St Margaret Newfifhft	114	66	St Michael Crookedla.	179	133
St Alhall-wes Breadft	35	16	St Dunftans Eaft	265	150	St Margaret Pattons	49	24	St Michael Queenha	203	122
St Alhallowes Great	455	426	St Edmunds Lumbard	70	20	St Mary Abchurch	99	54	St Michael Que ne	44	18
St Alhallowes Honilii	10	5	St Ethelborough	195	105	St Mary Aldermanbury	181	109	St Michael Royall	152	116
St Alhallowes Leffe	139	171	St Faiths	104	72	St Mary Aldermary	105	75	St Michael Woodftreet	112	62
St Alhall-Lumbardftr	90	55	St Fofters	144	105	St Mary le Bow	64	36	St Mildred Breadftreet	59	40
St Alhallowes Staining	185	112	St Gabriel Fen-church	69	39	St Mary Bothaw	55	50	St Mildred Poultrey	58	46
St Alhallowes the Wall	500	356	St George Botolphlane	41	27	St Mary Colechurch	17	6	St Nicholas Acons	46	28
St Alphage	271	133	St Gregories by Pauls	376	232	St Mary Hill	94	64	St Nicholas Coleabby	125	91
St Andrew Hubbard	71	25	St Helens	108	75	St Mary Mounthaw	56	37	St Nicholas Olaues	90	62
St Andrew Vndefhaft	274	189	St James Dukes place	162	190	St Mary Summerfet	342	243	St Olaues Hartftreet	237	160
St Andrew Windrobe	476	301	St James Garlickhithe	189	118	St Mary Stayninge	47	27	St Olaues Jewry	54	32
St Anne Alderigate	282	197	St John Baptift	138	83	St Mary Woolchurch	65	33	St Olaues Siluerftreet	250	132
St Anne Blacke-Friers	652	467	St John Euangelift	9		St Mary Woolnoth	75	38	St Pancras Soperlane	20	15
St Antholins Parifh	58	33	St John Zacharie	85	54	St Martins Ironmonger.	21	17	St Peters Cheape	61	35
St Auftins Parifh	43	10	St Katharine Coleman	299	213	St Martins Ludgate	196	128	St Peters Cornehill	136	76
St Barthol. Exchange	73	51	St Katherine Creech.	335	231	St Martins Orgars	110	71	St Peters Pauls Wharfe	114	86
St Bennet Fynch	47	21	St Lawrence Jewry	94	48	St Martins Outwitch	60	34	St Peters Poore	79	47
St Benn. Grace-church	57	41	St Lawrence Pountney	214	140	St Martins Vintrey	417	349	St Stevens Colmanftr	160	391
St Bennet Pauls Wharf	355	172	St Leonard Eaftcheap	42	27	St Matthew Fridayftr.	24	6	St Stevens Walbrooke	34	17
St Bennet Sherehog	11	1	St Leonard Fofterlane	335	255	St Maudlins Milkftreet	44	22	St Swithins	93	56
St Botolph Billingfgate	83	50	St Magnus Parifh	103	60	St Maudlins Oldfifhftr.	176	121	St Thomas Apoftle	63	110
Chrifts Church	653	467	St Margaret Lothbury	100	66	St Michael Baffifhaw	353	164	Trinitie Parifh	115	79
St Chriftophers	60	17									

Buried in the 97 Parifhes within the walls, — 15207 *Whereof of the Plague* — 9887

	Buried	Pla.		Buried	Pla.		Buried	Pla.		Buried	Pla.
St Andrew Holborn	3958	2103	Bridewell Precinct	230	179	St Dunftans Weft	958	665	St Saviours Southwark	3226	2446
St Bartholmew Grea	493	344	St Botolph Alderfga.	997	755	St George Southwark	1613	1260	St Sepulchres Parifh	4509	2746
St Bartholmew Leffe	307	139	St Botolph Algate	4926	4051	St Gues Copplegate	8069	4838	St Thomas Southwark	475	371
St Bridget	2111	1427	St Botolph Bifhopfg	3464	2500	St Olaues Southwark	4793	2785	Trinity Minories	168	122
									At the Pefthouse	159	156

Buried in the 16 Parifhes without the Walls — 41351 *Whereof, of the Plague* — 28888

	Buried	Pla.		Buried	Pla.		Buried	Pla.
St Giles in the Fields	4457	3216	St Katherines Tower	956	601	St Magdalen Bermon	1943	1362
Hackney Parifh	232	132	Lambeth Parifh	798	537	St Mary Newington	1272	1004
St James Clarkenwell	1803	1377	St Leonard Shoreditch	2669	1949	St Mary Iflington	696	593

St Mary Whitechappel 4766 855
Redriff Parifh 304 210
Stepney Parifh 8598 5583

Buried in the 12 out-Parifhes, in Middlefex and Surrey — 18554 *Whereof, of the Plague* — 21420

	Buried	Pla.		Buried	Pla.
St Clement Danes	1969	1319	St Mary Sauoy	303	198
St Paul Covent Garden	408	261	St Margaret Weftminft.	4710	3742
St Martins in the Fields	4804	2883	*buried at the Pefthouse*	156	

Buried in the 5 Parifhes in the City and Liberties of Weftminfter — 12194
Whereof of the Plague — 8403

The Total of all the Chriftnings — 9967
The Total of all the Burials this year — 97306
Whereof, of the Plague — 68596

The Difeafes and Cafualties this year.

Abortive and Stilborne	617	Executed	21	Palfie	30
Aged	1545	Flox and Small Pox	655	Plague	68596
Ague and Feaver	5257	Found dead in ftreets, fields, &c.	20	Planner	6
Appoplex and Suddenly	116	French Pox	86	Plurifie	15
Bedrid	10	Frighted	23	Poyfoned	1
Blafted	5	Gout and Sciatica	27	Quinfie	35
Bleeding	16	Grief	46	Rickets	557
Bloody Flux, Scowring & Flux	185	Griping in the Guts	1288	Rifing of the Lights	397
Burnt and Scalded	8	Hangd & made away themfelves	7	Rupture	34
Calenture	3	Headmould fhot & Mould fallen	14	Scurvy	105
Cancer, Gangrene and Fiftula	56	Jaundies	110	Shingles and Swine pox	2
Canker, and Thrufh	111	Impofthume	227	Sores, Ulcers, broken and bruifed	
Childbed	625	Kild by feverall accidents	46	Limbs	82
Chrifomes and Infants	1258	Kings Evill	86	Spleen	14
Cold and Cough	68	Leprofie	2	Spotted Feaver and Purples	1929
Collick and Winde	134	Lethargy	14	Stopping of the ftomack	332
Confumption and Tiffick	4808	Livergrown	20	Stone and Strangury	98
Convulfion and Mother	2036	Meagrom and Headach	12	Surfet	1251
Diftracted	5	Meafles	7	Teeth and Worms	2614
Dropfie and Timpany	1478	Murthered and Shot	9	Vomiting	51
Drowned	50	Overlaid & Starved	45	VVenn	1

Chriftned { Males — 5114 ; Females — 4853 ; In all — 9967 }
Buried { Males — 48569 ; Females — 48737 ; In all — 97306 } Of the Plague — 68596

Increafed in the Burials in the 130 Parifhes and at the Peft-houfe this year — 79009
Increafed of the Plague in the 130 Parifhes and at the Peft-houfe this year — 68590

Figure 5.1 Annual bill of mortality for 1665. L2926.2. Used by permission of the Folger Shakespeare Library under a Creative Commons Attribution-ShareAlike 4.0 International License.

sermons, exhorting his congregation to thank God for zeros in the plague columns: "for these 12. moneths and above, I finde there nothing but Ciphers: Ah Lord, how unthankful are we for such a blessing! when thou might'st as justly as suddenly, turn our *Ciphers* into *Figures*."[109] Even Charles I kept up with the London bills during the civil wars, as parliamentary agents discovered when they intercepted two suspicious packages and discovered "no more nor no less, then the Bills of Mortality bound up in a bundle, of the whole years burials in *London*, &c. usually sent heretofore to the King."[110]

Demand for both historical and current bills of mortality always increased dramatically during plague epidemics. Evidence suggests that plague-time demand for the bills resulted in unusually large print runs, though surviving copies are rare. During the 1660s, the Company of Parish Clerks lamented their inability to "Recover all the particular Weekly Bills thereof; the sight of them hath been much desired these times; but it is beyond my power, as yet, to answer mens expectations."[111] Rare book librarian Stephen Greenberg convincingly argued that the early bills of 1603 were printed using two different presses to increase the number of bills that could be printed during the single day the printer had to produce each week's bill. This would have resulted in a production run of somewhere between 5,000 and 6,000 bills per week in a city of 141,000, or one bill per 20 to 25 people.[112] Only at the end of the year and into 1604, when the worst of the epidemic was past, did the printer reduce his production to the approximately 2,500 bills that could be printed on a single press in one day.[113]

Londoners assiduously collected and evaluated the weekly mortality statistics printed in the bills over the course of plague epidemics. During the 1636 plague, one preacher noted how his parishioners "haue beene and are very diligent in enquiring after the weekly Bils of mortality, and they that could first obtaine the Bill from their Parish Clarks, haue acknowledged to be most beholden vnto them."[114] An anonymous Londoner faithfully filled out a comparative chart of weekly death tolls, subdivided by geographical area, in a printed broadside that listed remedies for the plague and exhorted its readers to "Live well. Die well." Merchants such as Edward Wood, who lived in Littleton but had business dealings in London, followed the bills just as carefully and requested their local factors send them copies with their letters.[115] Others acquired the same information by subscribing to newspapers such as the *London Gazette*, which printed summaries of the bills of mortality on the back of their 1665–6 broadsides.[116] The newspapers were also a source of provincial statistics, so that people could compare the extent of the infection in various localities; many English towns instituted their own bills of mortality in imitation of the London bills, and by 1665 the London newspapers could confidently report news about the plague in Bristol, Cambridge, Gosport, Norwich, and Yarmouth.[117] Even Daniel Defoe, in his fictional account of the 1665 plague, maintained a running commentary on the bills of mortality.[118]

In one of the most famous diaries of the seventeenth century, Samuel Pepys recorded a weekly summary of the bills of mortality from June, 1665 until the following May. His diary is invaluable for providing a first-hand account of the passion for and popular discussion of plague numbers during a London epidemic. Even before Pepys began tracking the bills, their numbers were the talk of the city.[119] By August, the plague had grown so dangerous that during a ride through London's suburbs, Pepys found

> all the way, people, Citizens, walking to and again to enquire how the plague is in the City this week by the Bill—which by chance at Greenwich I had heard was 2020 of the plague, and 3000 of all diseases; but methought it was a sad question to be so often asked me.[120]

A week later, at his office, Pepys and his coworkers were unable to concentrate on work and instead sat around discussing the plague numbers, "in great trouble to see the Bill this week rise so high."[121] Even on the official day of Thanksgiving to celebrate the end of the epidemic, in November of 1666, the numbers in the bills were on people's minds: eight people had died of the plague that previous week, and seven more would die in the week to come. Pepys lamented the rush to celebrate: "Lord, how the town doth say that it is hastened before the plague is quite over, there dying some people still."[122] There was thus a general consensus among the population of seventeenth-century London that the bills of mortality provided valuable, quantitative information on the severity of a plague epidemic.

This consensus included those who defended God's providential control of epidemic disease and saw plague death counts as a message from God. James Balmford, during the 1603 epidemic, equated God's love with a decreasing number of deaths reported by the bills and exhorted his congregation to "coole not in your deuotion, because the number of the buried in our parish is fallen (blessed by God) from 305. to 51. in one weeke, and from 57. to 4. buried in one day. Shall our loue coole, when Gods loue is kindled?"[123] An anonymous preacher, three decades later, complained that his parishioners had not "humbled themselues in Prayer, endeauouring to depart from their sinnes" when God made his wrath apparent via an increasing number of deaths.[124] During the 1636 epidemic, printed broadsides containing prayers for God's mercy were surrounded by death counts pulled from the bills of mortality, quantitatively reinforcing both the extent of God's wrath and the direness of London's situation. In 1665, clergyman Matthew Mead wished "we had Weeklie Bills of such Sins" as brought in the plague and used the language of quantified chance to predict it was "great odds, but the Contagion may shortly reach" the unrepentant sinner.[125] Regular fast days and days of Thanksgiving were observed throughout all the seventeenth-century epidemics, and death tolls used as proof of their effectiveness in appeasing God's wrath: "the two weeks Bills of Mortality, immediately after the Fast upon that occasion, were brought to the half of what they were the week before, and did amount not to more discernably then in the healthiest times."[126] Proof of God's providence was in the numbers.

Of all the information provided by the bills of mortality, increases in the death toll came under the most scrutiny, and those people who responded with fear or flight equated increasing death tolls with an increased risk of death. One preacher noted how increases in the bills frightened his parishioners, "made them murmure, and proiect to flee to their Country-houses here or there, and peraduenture to send beforehand their Wiues, Children, and Houshold-stuffe."[127] Both Pepys' diary entries and Defoe's novel highlight the same trend. Early in the 1665 epidemic, Pepys noticed plague deaths had risen by 50% in one week, hitting "267—where is about 90 more then the last." He concluded that there was now an unacceptable level of risk for his family and responded by both putting "all my affairs in the world in good order" and "sending of my wife's bedding and things today to Woolwich, in order to her removal thither."[128] Later in the epidemic, when the bill first topped a thousand—and again, when the first plague death occurred in his parish—Pepys similarly interpreted the numbers as indicating an increased risk of his own death. His thoughts turned once more toward "setting some papers in order, the plague growing very raging and my apprehensions of it great."[129] While Defoe's fictional protagonist chose to stay in London and recklessly gallivanted about town, exploring plague pits out of curiosity, his narrator's elder brother swiftly "sent his wife and two children into Bedfordshire, and resolved to follow them" after he had set his business affairs in order. The brother stayed in

London for several weeks, hoping to convince the narrator to join them, but eventually the risk of contracting the plague grew too high to allow further delay; "the bills were risen to almost 700 a week, and my brother told me he would venture to stay no longer."[130]

Indeed, people relied so heavily on the bills of mortality to evaluate the severity of plague outbreaks that there was real incentive to minimize death tolls by underreporting. At the start of the 1665 epidemic, the aldermen of Norwich attempted to hide the outbreak of plague by restricting local bills and falsifying the numbers reported to London papers.[131] Even during the height of the epidemic, when there was no hiding that plague deaths were occurring, not all clerks were completely honest about their numbers. Pepys recorded his concerns over underreporting after a chance meeting in the street with his parish clerk

> who upon my asking how the plague goes, he told me it encreases much, and much in our parish: "For," says he, "there died nine this week, though I have returned but six"—which is a very ill practice, and makes me think it is so in other places, and therefore the plague is much greater then people take it to be.[132]

Pepys was thus concerned that underreporting would cause people to incorrectly evaluate the severity of the epidemic, which would have particularly ill effects for people who based their risk-aversion decisions on the bills.

However, clerks were not the only ones responsible for reporting plague deaths and suspicion also fell on the searchers—usually poor women—who inspected dead bodies for "tokens" of plague and helped determine official causes of death for the parish register.[133] John Graunt, in his analysis of plague deaths, calculated that the plague was often underreported by "as many as one to four, there being a fourth part more dead of other casualties that year, then the years preceding or subsequent," and he blamed "the poor Searchers, out of ignorance, respect, love of money, or malice" for returning false verdicts.[134] Such accusations may owe more to elitism or misogyny than practice, as he later allowed that searchers were accurate in their assessment of other diseases. Searchers' testimony regarding causes of death was even considered credible evidence in late seventeenth- and early eighteenth-century murder trials.[135]

Regardless of the cause of the shortfall, Graunt developed a simple, mathematical solution: he looked at "the number that died of other diseases, and the casualties the weeks immedately before the Plague begun," and subtracted that number from the current totals. The remainder "are indeed dead of the Plague, though returned under the notion of those other diseases."[136] The practice of looking at so-called excess mortality to identify undercounting of deaths from infectious disease continues to the present day, albeit usually on the basis of annual rather than weekly statistics because of seasonal mortality spikes. The numbers in the bills of mortality were important enough to lie about during an epidemic, and even afterward they remained important enough to try to correct.

Over the course of the sixteenth century, widespread familiarity with gaming enabled the development of probabilistic ideas about chance and the creation of a new word, "odds," to express those ideas. While the more puritan clergymen wrangled with the theological implications of a chance based on natural laws, particularly its relationship to God's providence, others drew on the doctrine of secondary causes to reconcile chance and providence. Popular familiarity with gaming odds, combined with a belief in the existence of natural laws, led to the idea that the likelihood of all future events— and not just the roll of a die—could be quantified.

In the seventeenth century, this quantified chance proved particularly useful for both predicting and thus averting risk in people's daily lives. The increasingly widespread adoption of life and fire insurance reflected popular acceptance of the idea that risk could be assigned odds and monetary values—even if people weren't entirely clear yet on how to calculate those odds. This idea was not limited to those who had the resources to pay insurance premiums, as evidenced by the popular interest in mortality statistics, particularly the London Bills of Mortality. During plague years, people used the quantitative information in the bills to evaluate risk and even those who urged people to put their faith in God used the bills to support their arguments.

It is, however, important not to overstate the mathematical facility of the population of early modern England. At the height of the 1665 plague epidemic, Samuel Pepys heard a sermon preached by the Duke of Albemarle's chaplain, who chastised the bill-readers by proclaiming that "All our physicians can't tell what an ague is, and all our Arithmetique is not able to number the days of man." Only God's providence could determine the future, thus all the work of human physicians or mathematicians had been for naught. Pepys, listening to this sermon, agreed with the chaplain's conclusion, but not with the reasons behind it: "God knows, [it] is not the fault of arithmetique, but that our understandings reach not that thing."[137] Even though mathematicians would only develop mathematically rigorous theories of probability and statistics in the late seventeenth and eighteenth centuries, the men and women of early modern England had already embraced the concept of quantified chance.

6

"Davids Arithmetick":
Quantifying the People

In 1676, Lord Treasurer Thomas Osborne organized a census of English adults in order to determine "what proportion or disproportion of number there is" between Catholics and Protestants, and between Protestants who did and did not conform to the Church of England.[1] The Compton Census was run by and named after the bishop of London, who conducted it through established ecclesiastical data collection channels. Information was gathered at the parish level by rectors, victors, and churchwardens, then collated and summed up at the diocesan and archidiaconal levels.[2] While the local mechanisms of data collection were long established, the questions asked by the census were unfamiliar, leading to a fair amount of confusion about who was actually eligible to be counted. Approximately 20% of the parish officials who conducted the ground-level census carefully included detailed descriptions of the categories of people they had chosen to count. The rector of Knowlton provided a typical explanation of his numbers: "If by Persons inhabiting be meant Housholders, the Families are two; but otherwise of grown Persons, & such as be of yeers to come to the Sacraments of the Lords Supper, there may be sixteen."[3] While some returns only contained estimates of the parish population, others went to the opposite extreme and listed all the parish's inhabitants—adult or child—by name, rank, and household. Particularly enthusiastic parish officials, such as those in St. Nicholas's Warwick, even compiled "an exact account, in all respects, according to the best information we can get, by going to every house in the parish for our information herein."[4]

The Compton Census was not a neutral attempt to collect demographic data on England's adult population. Rather, it had a specific policy goal of reassuring Charles II there was no truth to the allegations that Catholics and Dissenters dramatically outnumbered the English men and women who conformed to the Church of England.[5] Despite Osborne's intentions, popular interest in the Compton Census numbers ended up becoming stronger

By the Numbers. Jessica Marie Otis, Oxford University Press. © Oxford University Press 2024.
DOI: 10.1093/oso/9780197608777.003.0007

than the interest of the government that had sponsored it. The census failed to have any substantial impact on monarchical policy, even though its results were presented to Charles II, James II, and William III. However, information from the census was cited in religious treatises, calculated and recalculated by political arithmeticians, and even published in a Dutch newspaper.[6] Instead of becoming a tool for monarchical policymaking, the data were embraced by political outsiders and used to discuss important religious and political issues in terms of numbers.[7]

By the early modern period, censuses were a familiar and long-standing tool of good government. However, the biblical precedents of King David and the association of censuses and taxation led to persistent, widespread reluctance to support anything that resembled "Davids arithmetic" or "numbering" people.[8] Locally controlled census data, such as the parish registers and censuses of the poor, largely avoided this association and became increasingly important to local governments over the late sixteenth and seventeenth centuries. The parish registers, in particular, became the foundation for the collection and analysis of plague mortality statistics, which in turn paved the way for more general seventeenth-century attempts to use demographic data for social analysis. This new way of thinking with numbers led to the post-Restoration articulation of a formal practice of political arithmetic that was designed to support monarchical policymaking. However, by the end of the century, proponents of political arithmetic had redefined their work as the analysis of neutral demographic data, divorced from any particular political agenda. This led to the transformation of political arithmetic and demographic statistics from a specific tool of government policymaking to a general method for using numbers to understand the population and society of early modern England.

Numbering the People

Censuses were an ancient tool of government, referenced in both classical history and the Bible and thus widely known among the people of early modern England. Men who attended grammar school learned of the ancient censors of Rome, who were employed to perform "a generall survey and numbring of the people" and whose census results "were reckoned and entred into the Censours bookes of cittizens."[9] The census-takers of the Bible appeared in biblical readings and sermons, including the backstory of

Jesus's Bethlehem birth and the stories of Moses, Saul, and David. Indeed, Moses's censuses were such a prominent component of his story that the fourth book of the Old Testament "is called Numbers, because therein are related two severall numbrings of the people."[10] These censuses were only intended to count free, adult males eligible for military service, not the entire population. By Moses's second survey, the Israelites numbered "six hundred twentie foure thousand seuen hundred seuenty three . . . besides the women, slaues, old men, and youth vnder twentie yeres, which were at the least twice as many" but whose precise numbers were considered unimportant.[11]

During the early modern period, one of the most famous of these ancient censuses was conducted by the Old Testament king David and led to divine punishment in the form of a deadly plague. English clergymen considered the census one of David's three great sins: "his adulterie, murther, and numbering of the people."[12] David's affair with Bathsheba and murder of her husband, Uriah, were sins that needed little explanation, but the problems inherent in David's census were not so obvious. Attempts to explain David's sin ranged from his trespassing on the prerogatives of the Roman Empire to forgetting to "leuie the summe of halfe a sickle vpon euery one" counted. Some even argued—conveniently ignoring the precedents of Moses and Saul—that it was completely unlawful to number the people since God had promised they would "be *innumerable* as the starres of the skie, and the sand of the Sea: and therefore it belonged vnto God onely to number that, which was innumerable."[13]

However, most clergymen agreed that the fault lay not in the census itself but rather in David's sinful reasons for conducing the census. Hugh Latimer, bishop of Worcester, preached a sermon to Edward VI in which he explained that "it was not the numbrynge of the people that offended God, for a king may number his people, but he dyd it of a pride, of an elacion of mynde, not accordyng to Gods ordinaunce."[14] Sixty years later, Andrew Willet, the rector of Barley in Hertfordshire, was only slightly less harsh, arguing that David's sin lay in "entring into a needlesse action, wherof there was no cause, but onely Dauids curiositie."[15] Shortly before the Restoration, the non-conformist Anthony Burgess argued that David's sin was in seeking to conduct a census "out of vain and ambitious ends," while William Guild, a Church of Scotland minister, placed the blame on all three motivations: "curiosity, vain confidence and pride, which makes the action vicious and sinfull."[16] While conducting a census was a prerogative of sovereign princes, it was also an act that pious Christians should perform only when there was an imme-diate, practical need for its results.[17]

Monarchical censuses for taxation and military levies had a long history within England, including the 1086 Domesday book and the 1334 assess-ment that set perpetual tax rates for the "fifteenth and tenth."[18] During the first half of the sixteenth century, royal ministers such as Cardinal Wolsey and Thomas Cromwell oversaw a variety of censuses or surveys intended to assess or extract population resources for immediate fiscal and military purposes.[19] A 1522 survey of England's military capacity formed the basis for the loans of 1522–3 and would have been used to assess the 1525 Amicable Grant had it not fallen through.[20] Muster rolls included both the names of men who could be called out for the militia and the names of taxpayers who were required to furnish them with uniforms, arms, and other equipment, which turned into a flat tax of 70s. per man by the end of the century.[21] The 1535 *Valor Ecclesiasticus* assessed the values of church properties with an eye toward confiscation, as did the Chantry Commissions of the 1540s.[22] The word "census" itself had such a strong financial connotation in both its Latin and English usage that in 1538 Sir Thomas Elyot defined a census as "yerely reuenues. Also valuation of goodes. Also a subsidie, the numbring of the people."[23]

One particular census, associated with the granting of subsidies, became a repeated and familiar occurrence during the sixteenth and seventeenth centuries. Although there had been at least ten attempts at directly assessed taxes prior to the reign of Henry VIII, they were largely failures. The subsidy only acquired its early modern form during the war years of the 1510s, under the direction of Cardinal Wolsey. Subsidy commissioners were appointed to assess the wealth of the population, including both annual income and the value of movable goods. Taxpayers would swear under oath to the value of their assets, then pay a proportional—rather than fixed—amount of taxes.[24] These subsidy valuations were a thorough census of potential taxpayers as they certified "the names and surnames of euery parsone man and woman of thage of .xv. yeres or aboue" who was not on charity, as well those under fifteen with "landes. tenement rentes, fees, annuytes, offyces, or Corodyes, of the yerely value of .xx.s. or aboue, or hauyng goodes or Catalles, mouable, coyne, plate, Stocke or marchandyse . . . to the value of .xl.s. or aboue."[25] Subsidy valuations proved so useful that they later formed the basis for assigning a host of other taxes, such as church rates, scavenger rates, poor rates, militia rates, and plague rates.[26]

The use of subsidy valuations to set tax rates incentivized potential taxpayers to evade being listed or to misreport their assets. Subsidy
valuations were fairly accurate during the first half of the century, but Queen
Elizabeth allowed the elimination of the taxpayer oath in 1563, which led to
rampant underreporting and tax evasion. In the wealthy London wards of
Cordwainer and Broad Streets, the percentage of taxpayers assessed at £50 or
more dropped from 30% in 1563 to a mere 5.5% by 1598. At the same time,
the percentage of taxpayers assessed at £3 in goods—at that time the lowest
assessment level—rose from 20% to 40%.[27] By the 1580s, underreporting
was such a well-known problem with the subsidy valuations that the government attempted to divorce these valuations from other rates. In 1584,
the earl of Huntingdon ordered his militia commissioners "to have regarde
not to the favorable and easie taxacion sett downe in the subsydie booke but
what there levings are inded by reasonable construction." Two years later,
Yorkshire justices of the peace who proved reluctant to meet their military
obligations were even threatened with the accurate assessment of their subsidy valuations: "yt maye happen that the observacion of other lawes will
be required of them which will touche their purses more deeplie then this
thinge doth."[28] Despite almost yearly subsidies granted from 1581 until the
end of Queen Elizabeth's reign in 1603, the combined total of these post-oath
subsidies was still less than the subsidy revenues of the 1540s and 1550s.[29]

During the seventeenth century, the decline of subsidy revenues
encouraged Stuart monarchs to invent new ways to assess and tax the wealth
of the English population. Customs duties and the excise proved the most
durable. Quota taxes such as the ship tax proved the most reliable method
for raising a specific sum of money. However, the method of assigning quotas
was often perceived to be grossly unfair, particularly after the outbreak of the
English Civil Wars, when Royalists could claim that their quotas had been ar-
tificially inflated as punishment for their politics.[30] The financial demands of
mid-century warfare also led to the levying of income-graduated poll taxes
in 1641 and 1660, as well as the creation of an entirely new form of census-
based taxation: the hearth tax.[31] Since the hearth tax was based on a combination of property values, income, and number of hearths in each household,
it proved a far more reliable indicator of household wealth than taxpayers'
monetary self-reports alone. However, it also created a source of general
discontent, as it required people to let strangers into their homes to count
their hearths. This led to its 1696 replacement by a window tax that could be
assessed from the outside.[32]

The association of censuses and taxation continued throughout the seventeenth century, making financial self-interest the primary barrier to more expansive, frequent, and reliable national censuses. The government as a whole recognized the need for accurate information on its people. In 1619, the Virginia assembly passed a registration bill requiring annual quantitative reports on the colony's demographic data. Between 1623 and 1700, the English government conducted twenty-one censuses of its various colonies in the West Indies and North America.[33] At the same time, however, the members of England's Parliament were reluctant to enact measures that would significantly impact their own wealth and resisted efforts to conduct thorough and accurate censuses of the population of England itself.[34] During a 1657 debate over the need to make "equal and equitable" assessments, members of the interregnum Parliament stonewalled a new census, noting "that the chief magistrate should know men's estates was always avoided." They also implied that any attempt to conduct a census might bring the fate of David down upon England by claiming the project looked "something like the numbering of the people."[35] Numbering colonists was an acceptable process and yielded valuable information for their home government, but numbering taxpayers at home—particularly taxpayers with political power—met serious, self-interested resistance.

From Local Lists to Demographic Analysis

English elites may have been reluctant participants in national censuses intended to extract financial resources from them, but those same people had less hesitation when it came to conducting their own smaller-scale censuses. Like royal ministers, they understood the census to be an important tool for gathering information and more generally recognized the potential of listing and counting for assessing, controlling, and extracting resources from the people they governed.[36] Local religious and civic officeholders, such as churchwardens or city aldermen, had long kept account books, but the second half of the sixteenth century saw the increasing use of paper-based recordkeeping as part of the rise of what historian Peter Burke has dubbed the "paper state."[37] Particularly in urban areas, these local government positions came to be dominated by the same middling sort of men who were increasingly incentivized by commercial pressures to learn reading, writing, and Arabic numeral arithmetic for their own personal affairs. When faced with religious, economic, demographic, or epidemiological crises, these officials coped by turning to the same sort of inventories, accounts, and paper records that they used in their businesses and households.[38] As local governments became more "data driven," they moved to protect their paper records—keeping them under lock and key at home—but also shared their data on a need-to-know basis through information networks that stretched across the country to other localities and the central government in London.[39]

The most regular and widespread of these local censuses were the parish registers. In 1538, Henry VIII ordered parish officials to record every birth, marriage, and death that occurred within the Church of England. Rumors immediately began to circulate, speculating that the registers were intended to form the basis for some new tax, but Henry's chief minister Thomas Cromwell reassured local justices of the peace that the registers were intended only to clarify lines of inheritance and allegiance.[40] While the registers were never intended to provide a quantitative account of the parish's population at any one moment in time, they did create an accurate and thorough census of people as they passed through critical life milestones. The registers were particularly significant in that they were not limited to the rich, taxpaying part of the population, nor to only the part consisting of able-bodied males fit for military service, but rather sought to collect information on the entire population; this inclusiveness would prove to be vital for the work of early demographers in the seventeenth century. Even in the sixteenth century, the value of the information contained in the parish registers was immediately apparent, and the registers themselves were closely guarded and maintained by parish officials across the country. Royal injunctions from 1561 required parish officials to send annual copies of their registers to their local diocese. These were followed up by proposed parliamentary bills in 1563 and 1576 aimed at creating a centralized collection point for the entire country. These bills failed to pass, and parish registers remained under the control of local officials and the Church of England except for a brief period during the Interregnum.[41] They did not even become a source for central government taxation until the passage of the Birth, Marriages, and Burials Duty Act of 1695.[42]

Another common type of local census that was conducted regularly, albeit not on an annual basis, was the census of the poor. In some ways the opposite of the subsidy valuations, these censuses were intended to identify those people who were supported on public funds or otherwise indigent and in potential need of charity. These generally took the form of a time-limited collection of data, rather than an ongoing collection effort like the parish registers.[43] Beginning in the 1550s and intensifying in the 1570s, urban and rural officeholders alike collected the names of their local poor.[44] The most famous of these was the 1570 Norwich Census of the Poor, which followed in the wake of an attempted uprising in the city and led the city government to fear that the poor might join and strengthen any future rebellions.[45] Late sixteenth-century demographic, political, and economic stresses—a booming population, war with Spain, and year after year of famines in the 1590s—led to increased population mobility and made it a matter of some urgency for urban officials to enumerate the poor, suppress vagrancy, and maintain the established social order. In 1598 and again in 1601, the parish became the local government unit responsible for implementing the new Elizabethan Poor Laws, further incentivizing them to keep track of which poor people they were—and weren't—responsible for maintaining.[46]

During the sixteenth century, the information gathered through national and local censuses largely remained in qualitative list form. This allowed government officials to deal with people on an individual basis—such as knowing who a child's parents were, what a particular taxpayer owed during a subsidy, and which local men and women were permanently unable to work because of infirmity or age. Data were sometimes stripped of identifying information and aggregated into a quantified, arithmetically manipulable format—notably, Wolsey's 1520s corn surveys and the 1570 Norwich Census of the Poor—but this was not common. Furthermore, any sums or analyses were generally limited to immediate purposes rather than used as part of a large-scale quantitative analysis of government population data.[47] However, one type of census data was regularly extracted from lists and put into a quantified—and mathematically manipulable—form: deaths from plague.

English's first plague mortality statistics were compiled around 1520, with surviving London examples dating back to at least 1528.[48] After the institution of parish registers, these numbers could be derived from the plague deaths listed in the registers at the instigation of city governments or, in the case of London, a combination of city and central government officials. Plague regulations from the 1578 epidemic required local authorities to count their dead, which led the city of Norwich to generate weekly mortality reports from 1579 until 1646; these are the best and longest-running series of early modern English vital statistics outside of London.[49] During the second half of the sixteenth century and through the seventeenth century, plague mortality statistics were compiled in

London, Norwich, Bristol, Chester, Exeter, Newcastle, Oxford, York, and other major cities.[50] Local government officials used these figures to inform their decision-making on issues such as quarantining travelers and goods from other cities, closing the city gates against outsiders altogether, and instituting plague rates to support those shut up in infected houses. Although these data were useful at a local level, the limited geographical scope of each set of mortality statistics meant they couldn't speak to epidemiological conditions on a country-wide level.

Intriguingly, despite the close association of David's biblical census with plague, there seems to be no indication that people saw a causal relationship between cities numbering their dead and the outbreak of plague. Even I. D.'s 1636 *Salomon's Pest-House*, which bewailed human sins and chastised people for putting their faith in numbers instead of God, failed to draw any general conclusions about censuses of the dead. Instead, I. D. discussed David's plague solely in terms of magisterial flight, noting "that plague [was] caused by his sinne, the numbring of the people, which caused such a sorrow in *Dauid*, that he was ready by his owne death to redeem the publike calamitie." I. D. then entreated magistrates to emulate David in staying to work for the good of their plague-stricken cities: "let not Magistrate forsake his Citie, nor the Minister his flocke" during an epidemic.[51] Yet nowhere did I. D. imply that these same magistrates were responsible for their plague outbreaks, in the same sense that David had been. Compiled mortality statistics were an effect, not a cause, of plague.

By the turn of the seventeenth century, calculating government officials and elites began to move beyond local and immediate analyses of lists and their sums to ask and attempt to answer more abstract policy question of their data. In 1601, Thomas Wilson tried to derive the total number of English subjects from the muster rolls of 1588, while in 1603, the Dutch Stranger and prominent London merchant Gerard Malynes made his own attempt by multiplying the size of the average family by the number of families in England—the latter two numbers supplied to him by "the obseruation of *Polititians*."[52] In 1607, a Privy Council paper used subsidy and muster rolls to calculate the wealth and population size of Somerset and Northamptonshire. The paper then attempted to justify the enclosure of pasture lands by pointing out the greater wealth and population of enclosed Somerset, as opposed to unenclosed Northamptonshire. In 1613, solicitor general and future lord chancellor Francis Bacon—a strong proponent of quantification— pointed out the analytic potential in the information that the government already gathered. "The greatness of a state in bulk or territory doth fall under measure; and the greatness of finances and revenue doth fall under computation. The population may appear by musters, and the number of cities and towns by charts and maps."[53] The *Oxford English Dictionary* even credits Bacon with the earliest use of the English term "population" in the sense of the inhabitants of a kingdom or other geographical area, which had become itself a unit of analysis.[54]

While people collected and analyzed information from a variety of sources—such as Andrew Willet's 1614 analysis of charitable bequests over time using livery company and prerogative court records[55]—the widespread availability of the printed London Bills of Mortality from 1603 onward made them a particularly popular source of demographic data in the seventeenth century. As early as 1607, baptism and burial numbers in the bills were cited as evidence of the city's increasing population.[56] During the first few decades of the century, much analysis of the bills was focused on active plague outbreaks but during the relatively plague-free years between 1636 and 1665, people began to increasingly analyze the bills to support a wider variety of arguments. An anonymous 1644 petition of "some hundreds of retaylers" argued that the abolition of the farthing as a coin of the realm had led to an increase in starvation among the poor.[57] Later that decade, London draper Rice Bush's treatise on caring for the poor advocated the establishment of food storehouses—not unlike modern food banks and food pantries—to feed the indigent. The proof of the effectiveness of these storehouses would be seen in the decrease in deaths by starvation as, ideally, the storehouses would altogether "prevent the mention of that sad disease in the weekly bills of mortality."[58] During the 1650s, James Howell used the bills to compare mortality rates between London and Amsterdam, reasoning that they would be proportional to the size of the respective cities' populations. Amsterdam's Bills of Mortality "at the utmost, come but to about threescore a week; whence may be inferred, that *London* is five times more populous; for the number that dies in Her every week, comes commonly, to near upon three hundred."[59] The argumentative power derived from analyzing the Bills of Mortality was even great enough for people to lament the limitations of the bills as a data source; Richard Whitlock wished for a rural version of the bills, while Thomas Culpepper complained that the recorded causes of death did not properly account for deaths from overindulgence in food and drink.[60]

These writers' analyses of demographic data were not mathematically sophisticated; anyone who could read Arabic numerals enough to

transfer the data into their preferred system for calculation could probably replicate what earlier calculators had done. These writers' methods
and results thus remained legible to the majority of the early modern
English population. Physician Marchamont Nedham used just addition
and subtraction to analyze the change in disease over time and argue
that his colleagues should no longer rely on ancient writers, regardless
of what established authorities such as the Royal College of Physicians
advocated. To prove his point he chose to examine a small subset of
diseases presented in the Bills of Mortality, "and leave you to observe and
reason out the rest in others" using the same methods.[61] After examining
the historical symptoms of scurvy, he turned to the bills to provide evidence of the disease's increase:

> if you look back on former dayes about 30 yeares agoe, the number reckoned
> to dye of that Disease was but small, and year after year ever since it hath
> encreased gradually. In the year 1630. the number was but 5. In 1631. 7 —
> 1632. 9. — 1636. 25. Afterward, in the year 1647. the Account came up to
> 32. — In the year 1652. to 43. — In 1655. to 44. — In 1656. to 103 . . . which
> signifieth, that the Disease is very much alter'd, being grown to a higher
> pitch of Malignity and Mortality than in former time.[62]

Nedham similarly analyzed consumption, convulsion, gout, measles, rheu-
matism, rickets, "Rising of the Lights," smallpox, and "Stopping of the
Stomach," using only the increase or decrease of annual mortality statistics
to argue his points.[63]

In 1662, John Graunt took the analysis of demographic data to a new level
with his *Natural and Political Observations*. Despite the breadth and scale
of his analyses, he derived them from the Bills of Mortality with only "the
Mathematiques of my Shop-Arithmetique," including addition, subtraction, multiplication, and ratios. With this, he "reduced several great confused *Volumes* [of bills] into a few perspicuous *Tables*, and abridged such
Observations as naturally flowed from them."[64] These observations were
demographic and political in nature, ranging from "not above one in four
thousand are *Starved*" and "there are about six Millions, and an half of people
in *England* and *Wales*" to "The City removes *Westwards*" and "the most
healthfull years are also the most *fruitfull*."[65]

In his treatise, Graunt also explained to potential critics "to what purpose
tends all this laborious buzzling, and groping" with numbers.[66] He argued

that the purpose of good government was to provide peace and prosperity to its subjects, and the foundation of this was a knowledge of both land and people: "It were good to know the *Geometrical* Content, Figure, and Scituation of all the Lands of the Kingdom. . . . It is no less necessary to know how many People there be of each Sex, State, Age, Religion, Trade, Rank, or Degree."[67] This knowledge would enable the government to estimate the needs of human consumption and make trade more regular, to the profit of the country as a whole. Furthermore, knowing "how small a part of the People work upon necessary Labours, and Callings," as well as the distribution of trades, "is necessary in order to good, certain, and easie Government, and even to balance Parties, and factions both in *Church* and *State*." Indeed, Graunt considered this information so important that he concluded his treatise by wondering if he had done the right thing in publishing rather than reserving his analyses for government use only.[68] His printers were less ambivalent about making this information publicly available, given there was robust enough demand for it to be reprinted three times in under three years.[69]

By the end of the seventeenth century, while the original meaning of David's census was not forgotten, public interest in demographic data had grown strong enough that judge Sir Matthew Hale could analyze David's census in a wholly demographic context:

> in the latter end of the Reign of *David*, about the Year of the World 2925, which was 435 Years after the Numbring of the People by *Moses* and *Eleazar*, *David* again Numbers the People, and then the Account of the People . . . [was] in all 1300000 fighting Men: and if we should take in Women, Children, and Aged, it is probable they were above five Millions. So that in the space of 435 Years, notwithstanding all these Decrements they were increased about three Millions.[70]

For Hale, the significance of David's census was not David's sin and the plague that followed. Instead, Hale focused on the numbers that resulted from each of the biblical censuses and compared them to one another to argue that, plagues and hardships notwithstanding, the overall population of Israelites had dramatically increased between the exodus from Egypt and the reign of David. Numbering the people was no longer an act of pride but a source of data that supported good government even without an immediate fiscal or military purpose.

Political Arithmetic

The growing association of demographic data analysis and good government led to the late seventeenth-century development of a practice called political arithmetic. The term itself originated no later than the 1670s in the writings of William Petty, an English government official in colonial Ireland with a background in natural philosophy.[71] Like Graunt, Petty looked to numbers as a method for making demographic arguments and deliberately contrasted "using only Comparative and Superlative words and Intellectual Arguments" with "the course (as a Specimen of the Political Arithmetick I have long aimed at) to express my self in Number, Weight, and Measure."[72] While they never officially co-authored a publication, Petty supplied Graunt with some of the data used in his analyses, and Petty's later work on the Dublin Bills of Mortality bore such a striking resemblance to Graunt's that some of their successors mistakenly attributed Graunt's work to Petty.[73] But unlike earlier demographers, Petty treated populations not only as things that could be described and made legible through quantitative means but also as things that could subsequently be arithmetically manipulated. Crucially, the ultimate purpose of Petty's political arithmetic was the advancement of the current monarch's policies—whatever they might be—and his analysis and manipulation of demographic data were simply effective means to that end.[74]

Petty's political arithmetic transformed political problems into demographic ones and reframed the work of the government as the manipulation of populations, particularly the ratio of one subgroup to another.[75] In 1670, Petty calculated ratios for the population of lands he held in Ireland, where "the right numbering of the People [demonstrated] that there were living about then, upwards of 500 Papists for one of these Protestants."[76] The 1676 Compton Census similarly reduced the religious populations of England to such ratios by determining that only one person out of twenty-two was nonconformist and one out of 178 a Catholic.[77] Once known, these demographic ratios could subsequently be manipulated by the transplantation of people; this was both the key to Petty's political arithmetic and a source of its failure in practice, as demographic manipulation required an enormous movement of people over space. One of Petty's plans involved exchanging 200,000 Irish men, women, and children with a like number of English ones; the former would constitute less than 2% of the new population of England, while the latter would definitively shift the balance of Protestants to Catholics in Ireland. Other, similar proposals included the exchange of Irish for English women, or even the replacement of Irish Catholic priests with English ones, thus getting rid of troublesome English Catholics while simultaneously destroying Catholicism's relationship to Irish nationalism.[78] Although developed to solve specific governmental problems within the context of Ireland, Petty's political arithmetic evolved after the accession of James II to become a more generalized method of social engineering that could—in theory, if never in practice—be deployed to manipulate and control populations throughout the Stuart composite monarchy and expanding British empire.[79]

Like the work of other seventeenth-century writers, Petty's manipulation of ratios might have been quantitative and arithmetical but did not require extensive mathematical knowledge to understand. In fact, Graunt's earlier manipulation of statistics in his *Observations* was more complicated than Petty's works.[80] During a presentation to the Royal Society, Petty attempted "to excite the World to the study of a little Mathematicks, by shewing the use of *Duplicate Proportions* in some of the most weighty of Humane affairs." However, he reassured his listeners that the mathematics of proportion should be simple enough that "the meanest Member of adult Mankind is capable of understanding" and "a Child of 12 years old may learn in an hour."[81] The presentation was well received among the members of the Royal Society, but Petty had overestimated the mathematical learning of some of his fellow government officials. Upon reading the printed version of the presentation, the solicitor-general of Ireland was particularly angry with Petty "for saying a Child of 12 yeares old might do in one houre what he found he could not do in many."[82]

Although Petty's arguments were expressed quantitatively and his program relied on the arithmetical manipulation of various ratios, the fact that such manipulations were a means to an externally determined end meant that Petty was able to tolerate a large degree of uncertainty and inaccuracy in his demographic data. He defended the use of numbers even in situations where he could not acquire exact information, noting that even those "that ayme at so much cleernes & certainty are forced sometimes to fly to estimate & opinion."[83] The important thing was to make certain that

the Observation or Positions expressed by *Number, Weight,* and *Measure,* upon which I bottom the ensuing Discourses, are either true, or not apparently false... and if they are false, not so false as to destroy the Argument they are brought for; but at worst are sufficient as Suppositions to shew the way to that Knowledge I aim at.[84]

For example, in his second set of observations on the Dublin Bills of Mortality, Petty noted an improbably large increase in the number of Dublin houses. Upon comparing the alleged numbers of houses, hearths, and burials, he concluded it likely to have been an error then fell back on experience, rather than mathematics, to claim "there is no ground from experience to think that in 11 year, the Houses in *Dublin* have encreased from 3850 to 6025. Moreover, I rather think that the number of 6025 is yet short." However, he was still willing to use the numbers as they were "the best estimate I can make of that matter, which I hope Authority will ere long rectifie, by direct and exact Enquiries."[85] Indeed, Petty repeatedly called for more thorough and exact surveys of the English and Irish populations, and English lands, in order to support government policymaking and the practice of political arithmetic.[86] The most important thing was not the absolute truth or falsity of his demographic data but rather that the numbers—and the immediate conclusions drawn from those numbers—were not so obviously false as to destroy the political arguments they were intended to support.

After Petty's death, subsequent political arithmeticians such as Gregory King and Charles Davenant began to disassociate their still-emerging practices from Petty's policy- and transmutation-centric version of political arithmetic. They acknowledged Petty's founding role in the field, noting Petty "first gave it that Name, and brought it into Rules and Method," namely, that "Mathematical Reasoning, is not only applicable to Lines and Numbers, but affords the best means of Judging in all concerns of humane life."[87] But they also focused on his printed works to the exclusion of his manuscript ones and preferred to treat populations as more neutral objects of knowledge, subject to natural laws that could be examined through quantitative analysis.[88] In 1691, Royal Society member John Houghton took it as given that "an account of the several Parts of the *Bills of Mortality* of *London*" would be of interest to "the *Political Arithmeticians* that desire to know the *Increase* and *Decrease* of Places," but he ascribed no particular agenda to his intended audience.[89] Throughout the 1690s, King argued that it was the job of the political arithmetician to draw conclusions from the numbers, not manipulate the numbers to support a predetermined conclusion or government policy. He agreed that "to be well apprized of the true state, and condition of a nation, especially in the two main articles, of its people, and wealth, [was] a piece of political knowledge." However, he castigated the exaggeration of "all former calculations" of English demographic data, whose untruths had the potential to lead government decision-makers astray.[90] While King was also forced to make educated guesses over the course of his calculations, he insisted that his numbers "come very near the truth." Any errors were not a deliberate manipulation meant to advance a specific political agenda but a natural result of the limitations of his original data.[91]

Like King, Davenant also argued for the neutrality of political arithmetic, casting it as a methodology with no inherent political agenda that could be applied to an ever-increasing set of political contexts. He defined the discipline as "the Art of Reasoning, by Figures, upon things relating to Government" and noted the "Foundation of this Art is to be laid in some competent Knowledge of the Numbers of the People."[92] It was not a method of government but rather a method of reasoning about government-related matters, whose foundations lay in numbers and demography. This method of reasoning provided quantitative evidence that might influence policy, but the evidence itself was politically neutral.

> A great Statesman, by consulting all sort of Men, and by contemplating the universal Posture of the Nation, its Power, Strength, Trade, Wealth and Revenues, in any Council he is to offer, by summing up the Difficulties on either Side, and by computing upon the whole, shall be able to form a sound Judgment, and to give a right Advice: And this is what we mean by Political Arithmetick.[93]

The political arithmetician calculated the difficulties of all possible courses of action and only then used these data to decide which course of action to advocate. Even then, this was no more than a judgment, "Political Arithmetick being a good Guide in these Matters; though it gives not demonstrative Proofs." At the end of the day, the primary purpose of political arithmetic's calculations was "So that the Parliament would not be quite in the Dark" when making policy decisions, not to make the policy decisions for them.[94] By the beginning of the eighteenth century, political arithmetic had thus become a method of producing evidence and arguments for parliamentary and public debates about monarchical policy, drawing on censuses and other quantitative data generated specifically for this purpose.[95] Davenant's arithmetic was very different from David's arithmetic.

Although the census was an ancient and well-known practice in early modern England, ideas about its acceptable uses and limitations were filtered through the biblical story of David, which warned against curiosity and idle quantification. Census data were supposed to be gathered and analyzed only for immediate fiscal or military purposes. The establishment of subsidy valuations under Henry VIII reinforced the popular association of censuses and taxation and led to self-interested resistance to and attempts to evade accurate reporting during central government census efforts. The primary exception to this was the parish registers, both because they were avowedly non-financial in nature and also because they remained under the local control of parish officials and ultimately answered to the Church of England rather than to civil officials.

However, from the mid-sixteenth century onward, local government offices came to be increasingly dominated by the same middling sort of men whose business interests incentivized them to learn to read, write, and calculate with Arabic numeral arithmetic. When faced with the myriad crises of the late sixteenth and seventeenth centuries, they turned to the information gathering tools they knew best, listing and increasingly summing and analyzing the people they governed. While many of these data were kept secret—or at least provided only on a need-to-know-basis—the one type of data that was freely shared was plague mortality statistics, with the city of London's Bills of Mortality forming an extreme example of public, quantified data. Over the first half of the seventeenth century, people increasingly quantitatively analyzed whatever demographic data they could acquire, particularly data from the London bills, and used the results to bolster their larger arguments about English society and politics.

By the end of the seventeenth century, the work of quantitatively inclined demographers, especially John Graunt and William Petty, led the latter to articulate a practice of governing by numbers called political arithmetic. In its original formulation, political arithmetic explicitly harnessed demographic analysis and the manipulation of population ratios to advance a monarch's policy goals in a composite kingdom and nascent colonial empire. After his death, Petty's successors reframed political arithmetic as a method of reasoning on demographics that could be applied to political matters but was inherently neutral and divorced from any specific policy goals. By the eighteenth century, political arithmetic was no longer a method of government. Indeed, it had transformed into a method of thinking and arguing numerically, that could be generally applied to whatever political, economic, and social questions faced the politically engaged men and women who increasingly chose to adopt it.

Epilogue

"Heau'ns Great Arithmetician": Living in a Numerical World

In 1599, playwright Thomas Dekker composed a comedy for the Lord Admiral's Men to perform before Queen Elizabeth and her court during that year's Christmas season. In the epilogue, an old man beseeched the audience to pray to God for the long continuation of Elizabeth's reign and framed his own prayer in numerical terms:

> And that heau'ns great Arithmetician,
> (Who in the Scales of Nomber weyes the world)
> May still to fortie two, and one yeere more,
> And stil adde one to one, that went before,
> And multiply fowre tennes by many a ten:
> To this I crie Amen.[1]

Dekker's play is just one illustration of the way English ideas about numbers were inextricably bound up with ideas about God in the early modern period. God was the "great Arithmetician" who created the world in number, weight, and measure. Afterward, God continued to use "the Scales of Nomber" to weigh the world and reckon the days of every man and woman's life. One of the many ways God showed favor toward Elizabeth was through acts of arithmetic, adding still more years to a reign that had already lasted forty-two years. In his enthusiasm, Dekker even suggested that God might multiply the decades and allow her to live as long as the prophets and kings of the Old Testament. God's love was in the numbers.

The people of early modern England developed a numerical world-view that existed within a framework defined by Christianity. Their world functioned according to quantitative laws because God had made it so, and people could learn these laws because God had given humans the ability

By the Numbers. Jessica Marie Otis, Oxford University Press. © Oxford University Press 2024.
DOI: 10.1093/oso/9780197608777.003.0008

to comprehend numbers when he raised them above all other creatures. Early modern attempts to quantify abstract phenomena like time and chance reflect this religious worldview. Calendar reform and concerns over the relativity of time were inspired by the need to celebrate religious festivals at specific chronological moments, while the impulse to quantify chance led to debates over the mechanisms of God's providence and how he directed earthly events through quantifiable laws. Even a proposal to conduct a national census was interpreted within the context of biblical censuses and God's directives to his chosen people. Early modern numbers were not value-neutral conveyors of quantitative information but rather the building blocks of Creation.

Numbers were a socially pervasive technology of knowledge, with the power to shape modes of thought. As much as possible, *By the Numbers* has reconstructed a cognitive element of English culture by looking at not just how people used numbers, but how people thought about numbers and how they used them to think about events in their day-to-day lives. In analyzing numbers this way, it gives ordinary people a stake in philosophies and technologies of knowledge, and in doing so reconnects some of the revolutionary changes in early modern science with the changes occurring in everyday numeracy. This period witnessed not only the mathematization of elite natural philosophy but also the increasing use of numbers by ordinary men and women to interpret the world around them. Many later eighteenth-century developments in economics and politics, among other fields, had antecedents in early modern ways of thinking about numbers. The men and women of seventeenth-century England understood the potential of numbers as a technology of knowledge, even if they had not yet developed the mathematics to completely explain the world via numbers.

It is critical to better understand numeracy as it was one of the most widespread technologies of knowledge that people used while attempting to impose order on the world around them. By looking at a diverse array of both material and textual sources, this book has demonstrated how the multiplicity of oral, object-based, and written early modern symbolic systems ensured that people across the socioeconomic and educational spectrums could perform arithmetical calculations and create permanent numerical records. During the sixteenth century, the combination of counting boards, tally sticks, and Roman numerals could serve the needs of the entire population, from the illiterate to highly educated political officeholders. Only at the turn of the seventeenth century did the literate subsection of the population begin to favor the use of Arabic numerals. Even then, this preference never operated to the complete exclusion of other systems.

The early modern adoption of Arabic numerals is particularly significant for modern historians because it linked the previously separate functions of permanent recording and calculation via the act of writing. The patterns of change in numerical symbol use and in the educational methods employed during the seventeenth century underscore the increasing conceptual connection between writing and arithmetic. By the eighteenth century, the dominant form of English numeracy was a self-consciously literary form of numeracy—the same numeracy that remains the default Anglophone conception of numeracy today. But in combination with an emphasis on written sources, this literary form of numeracy can obscure the vitality of different types of numeracy in the early modern and other periods. In one of the most pervasive examples, it leads modern scholars to make assumptions about the difficulty early modern people must have had performing calculations with Roman numerals. Yet someone in early modern England would no more have turned to Roman numerals to perform calculations than we would use a fork to eat soup.

Scholars from a variety of historical subfields have also used early modern numerical practices to make arguments about the advent of globe-spanning phenomena such as capitalism or modernity. Whether they are historians of science and mathematics looking to Arabic numerals and the explicit zero or business and economic historians interested in merchants' accounting and the use—or lack thereof—of double entry bookkeeping, these scholars have put quantitative changes at the heart of their grand narratives of history. It is therefore crucial to examine the full range of historical numeracies and to neither misunderstand the nature nor underestimate the pervasiveness of numerical concepts in the past.

Numbers are not transcendent phenomena. Acknowledging them as historically situated and looking at them through the lens of early modern English culture reveals major transformations in the way people used and thought about numbers during the sixteenth and seventeenth centuries. Changes in symbolic systems and educational practices encouraged new ways of thinking, as people began to consider numbers as a tool for interpreting common phenomena in their everyday lives. By the end of the seventeenth century, numbers were no longer simply a limited tool for financial and astronomical calculations. They had become the building blocks of

God's Creation and the foundation of English men and women's efforts to understand their world.

Over the next three centuries, the people of England—both those who stayed at home and those who joined colonial expeditions to conquer large swaths of the globe—increased their commitment to numbers as a technology of knowledge. Buildings began to be identified by numbers rather than signs, which laid the foundation for geographical locations, movable objects, and people themselves to be likewise assigned numeric identifiers. Complex fiscal calculations became normalized at multiple levels of society, informing household finances, global commerce, and international diplomacy. Personal insurance schemes proliferated based on increasingly mathematically rigorous probability calculations, particularly around human life expectancies. Statistical analyses came to undergird everything from institutional and government policymaking to polling and sports betting. Some people turned to the idea of a "clockwork universe" to reconcile a mathematical and mechanistic world with their notions of God. Over time, Anglophone men and women came to take for granted that they lived in a quantitative and mathematized world.

An inherently mathematical worldview encompasses both the mathematics of advanced scientific disciplines and the use of numbers in ordinary people's everyday lives. In the twenty-first century, most of the population of the Anglosphere is not highly skilled in mathematics by the standards of modern mathematicians. If asked, many people will self-deprecatingly refer to themselves as "bad at math," while others will lament the deplorable state of mathematical education in their local schools. At the same time we rely heavily on numbers in our everyday lives, using them to explain such diverse subjects as the location of our homes, our educational achievements, and whether our organs are functioning properly. We take for granted the opaque mathematical algorithms that set our car insurance rates or run internet search engines.[2] Ideas about numbers and numeracy are culturally determined, and they can vary significantly over time and place. But it is this essentially numerical way the general population understands the world that defines a mathematical worldview—both in the modern Anglosphere and in early modern England.

Arithmetic Textbook Marginalia by First Author and Holding Library

Table A.1 Marginalia by First Author

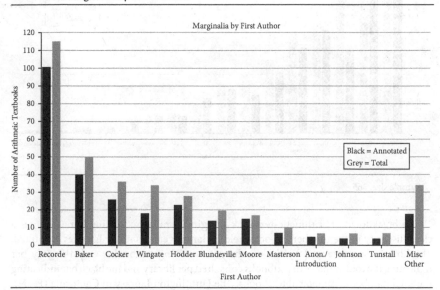

Marginalia by First Author

Sample size: 365

The breakdown of arithmetic textbook marginalia rates by first author—that is, the original author who gave his name to each textbook series—indicates some differences in the marginalia rates for each series. The authors are arranged by the number of surviving textbooks, with the gray bar indicating the total number of textbooks consulted per author and the black bar indicating the total number of annotated textbooks. Robert Recorde's textbooks are particularly highly annotated (101 out of 115 examined) and have the largest survival rate in the sample. The only author who equals this annotation rate is Moore (15 of 17), although Baker and Hodder come close. Baker's textbooks have the second-highest survival rate (40 out of 50 annotated), while the high number of surviving editions of Cocker's textbooks, published in the late seventeenth century, hint at his burgeoning popularity that continued into the eighteenth century. Wingate's arithmetics are outliers, with significantly less annotation (18 out of 34) than others.

Table A.2 Marginalia by Holding Library

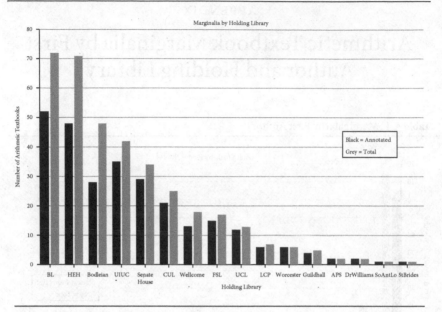

Sample size: 365

The breakdown of arithmetic-textbook marginalia rates by holding library shows a clear disparity in the number of books held by larger versus smaller libraries consulted. The libraries are arranged by the number of surviving textbooks, with the gray bar indicating the total number of textbooks consulted per library and the black bar indicating the total number of annotated textbooks. The Huntington Library in California (HEH), subject of William Sherman's in-depth study, has 66% annotated books in line with its nearest neighbors in size, the British Library (BL) at 72% and the Bodleian Library at the University of Oxford at 58%. Other libraries with smaller collections have even higher percentages of annotated books, which aligns with the tendency of larger libraries to have historically prioritized collecting "clean," unannotated copies of books as opposed to smaller libraries that obtained these books on a more ad hoc basis.

Notes

Abbreviations

APS	American Philosophical Society
BL	British Library
Bodl.	Bodleian Library
CUL	Cambridge University Library
ESTC	*English Short Title Catalogue*
FSL	Folger Shakespeare Library
HEH	Huntington Library
KJV	King James Version of the Bible
LMA	London Metropolitan Archives
OED	*Oxford English Dictionary*
SHL	Senate House Library
TNA	The National Archives of the UK
UIUC	University of Illinois, Urbana-Champaign
WL	Wellcome Library

Introduction

1. George Wharton, *Calendarium Ecclesiasticum: or, A New Almanack After the Old Fashion* (London: J. Grismond, 1659), C5r. The group of men whom I refer to with the catch-all title of "mathematicians" gave themselves a variety of titles, including well-wishers, lovers, students, and practitioners of mathematics. Those who authored almanacs, in particular, tended to style themselves as "Philomathematicus."

2. The almanac belonged to Henry Oxinden of Barham, who extensively annotated it, but it is unclear whether he or another member of his household was responsible for the underlining. HEH, 479104, f. D5r. For more on Oxinden, see *Oxford Dictionary of National Biography*, s.v. "Oxinden [Oxenden], Henry (1609–1670)," by Sheila Hingley, https://doi.org/10.1093/ref:odnb/21053.

3. Variations on this concept, without the Christian God, date back to at least the ancient Greeks. Plato's *Timaeus* describes a creator-god who uses geometry to impose order on the fabric of the world. John Fauvel and Jeremy Gray, eds., *The History of Mathematics: A Reader* (New York: Palgrave Macmillan, 1987), 76–9.

4. While "science" is arguably an anachronistic term when understood in the modern sense, it also appears frequently in early modern mathematical writings. I follow the usage pioneered by Deborah Harkness, who argues that the Elizabethan definition of

a science is "both a study of the natural world and a manipulation of the natural world for productive and profitable ends." Deborah Harkness, *The Jewel House: Elizabethan London and the Scientific Revolution* (New Haven, CT: Yale University Press, 2007), xv.

5. Robert Recorde, *The Ground of Artes Teachyng the Worke and Practise of Arithmetike* (London: R. Wolfe, 1543), A1v–A2r, A4v, A2v.

6. Another version of the adjective "numerate"—defined as "counted"—dates to at least the end of the fifteenth century, while the verb form—meaning "to count"—dates to the mid-seventeenth century. These were just two of a variety of number-related words that the *OED* first attests to in the early modern period, including numerated, numeration, numerical, and numerically. *OED*, s.vv. "numerate, *adj*.1," "numerate, *v*.," "numerated, *adj*.," "numeration, *n*.," "numerical, *adj*. and *n*.," and "numerically, *adv*." This is consistent with the growing interest in counting and quantification during the early modern period.

7. *OED*, s.vv. "numerate, *adj*.2," "numeracy, *n*."

8. *OED*, s.v. "literacy, *n*." The definition continues with this: "Also: the extent of this in a given community, region, period, etc. Cf. "NUMERACY *n*." This variant meaning and the cross-reference to numeracy are recent additions to the primary definition and were not there when I originally began this research. The previous definition, which more closely parallels the *OED* definition of "numeracy," was "the quality or state of being literate; knowledge of letters; condition in respect to education, *esp*. the ability to read and write." *OED* 2nd ed., s.v. "literacy," https://web.archive.org/web/20200712215613/https://oed.com/oed2/00134127. For comparison, the definition of "literate" as a noun is "A person who can read and write. Opposed to *illiterate*." *OED*, s.v. "literate, *adj*. and *n*."

9. For an extended discussion of the various types of early modern literacy, see Keith Thomas, "The Meaning of Literacy in Early Modern England," in *The Written Word: Literacy in Transition, Wolfson College Lectures 1985*, edited by Gerd Baumann (Oxford: Clarendon Press, 1986): 97–131.

10. *OED*, s.v. "numeracy, *n*."

11. Some modern scholars attempting to quantify historical numeracy rates have used age heaping as an identifiable behavioral marker that can stand as a proxy for innumeracy. See, for example, Brian A'Hearn, Jörg Baten, and Dorothee Crayen, "Quantifying Quantitative Literacy: Age Heaping and the History of Human Capital," *Journal of Economic History* 69, no. 3 (2009): 783–808. However, this does not work for defining numeracy as there is no necessary relationship between quantitative ability and a cultural belief in the importance of precisely and accurately reporting an adult's age in years.

12. *OED*, s.v. "numerate, *adj*.2."

13. While early modern authors almost always used the term "man" as opposed to "human," I have found no evidence that they believed women incapable of performing the same mathematical feats as men. Surviving account-books and almanacs demonstrate that early modern Englishwomen also performed bookkeeping for their households. In 1686 the Royal Society in Dublin examined the case of an eleven-year-old girl who "understands all Arithmetick, and Algebra, Trigonometrie, and the use

NOTES TO PAGES 4–6 165

of the Globes," and determined that there was nothing "extraordinary in her natare to Mathematick: but doe impute all to her timely earlie education." Bodl., MS Aubrey 10, ff. 10r, 29v.

14. Recorde, *The Ground of Artes* (1543), _6v.

15. Recorde, *The Ground of Artes* (1543), _6r.

16. Great apes, parrots, ravens, and rats can all subitize. My own informal experiments have suggested that my cats are also capable of subitizing when presented with piles of treats. While subitizing does not require counting, it does require active attention. Stanislas Dehaene, *The Number Sense: How the Mind Creates Mathematics* (New York: Oxford University Press, 2011), 259. Kim Uittenhove and Patrick Lemaire, "Numerical Cognition during Cognitive Aging," in Roi Cohen Kadosh and Ann Dowker, eds., *The Oxford Handbook of Numerical Cognition* (Oxford: Oxford University Press, 2015), 348.

17. Brian Butterworth, *What Counts: How Every Brain Is Hardwired for Math* (New York: Free Press, 1999), 226. Sarah Benson-Amram, Geoff Gilfillan, and Karen McComb, "Numerical Assessment in the Wild: Insights from Social Carnivores," *Philosophical Transactions of the Royal Society B: Biological Sciences* 373, no. 1740 (February 19, 2018). Christian Agrillo and Angelo Bisazza, "Understanding the Origin of Number Sense: A Review of Fish Studies," *Philosophical Transactions of the Royal Society B: Biological Sciences* 373, no. 1740 (February 19, 2018). Rosa Rugani, "Towards Numerical Cognition's Origin: Insights from Day-Old Domestic Chicks," *Philosophical Transactions of the Royal Society B: Biological Sciences* 373, no. 1740 (February 19, 2018). Oliver Lindemann and Martin H. Fischer, "Cognitive Foundations of Human Number Representations and Mental Arithmetic," in Roi Cohen Kadosh and Ann Dowker, eds., *The Oxford Handbook of Numerical Cognition* (Oxford: Oxford University Press, 2015), 36. Caleb Everett, *Numbers and the Making of Us: Counting and the Course of Human Cultures* (Cambridge, MA: Harvard University Press, 2017), 187. Dehaene, *The Number Sense*, 28.

18. John Bulwer, *Chirologia, or, The naturall language of the hand composed of the speaking motions, and discoursing gestures thereof: whereunto is added Chironomia, or, The art of manuall rhetoricke.* (London: Thomas Harper, 1644), 185.

19. Bulwer, *Chirologia*, 185.

20. Anthony Fitz-herbert, *The New Natura Brevium, Of the most Reverend Judge Mr. Anthony Fitz-Herbert* (London: W. Lee, M. Walbank, D. Pakeman, and G. Bedett, 1652), 583. This work was first printed in Latin circa 1534 and went through almost two dozen reprints in the sixteenth and seventeenth centuries.

21. For more on the consequences of being declared mentally incompetent in early modern England, see Richard Neugebauer, "Mental Handicap in Medieval and Early Modern England: Criteria, Measurement and Care," in Anne Digby and David Wright, eds., *From Idiocy to Mental Deficiency: Historical Perspectives on People with Learning Disabilities* (London: Routledge, 1996), 22–43.

22. By six months, babies can recognize differences with a 1:2 ratio, by ten months, they can recognize a 2:3 ratio, and by adulthood most humans can recognize somewhere between a 7:8 and 11:12 ratio at a glance. While adult capabilities are partially the

result of culture and formal education, the improvements in infants' abilities occur naturally as their brains mature. Christian Agrillo, "Numerical and Arithmetic Abilities in Non-primate Species," in Roi Cohen Kadosh and Ann Dowker, eds., *The Oxford Handbook of Numerical Cognition* (Oxford: Oxford University Press, 2015), 228. Manuela Piazza et al., "Learning to Focus on Number," *Cognition* 181 (December 1, 2018): 35–45. Titia Gebuis and Bert Reynvoet, "Number Representations and Their Relation with Mathematical Ability," in Roi Cohen Kadosh and Ann Dowker, eds., *The Oxford Handbook of Numerical Cognition* (Oxford: Oxford University Press, 2015), 331.

23. Koleen McCrink and Wesley Birdsall, "Numerical Abilities and Arithmetic in Infancy," in Roi Cohen Kadosh and Ann Dowker, eds. *The Oxford Handbook of Numerical Cognition* (Oxford: Oxford University Press, 2015), 263, 267.

24. Dehaene, *The Number Sense*, 28.

25. In Brazil, the Mundurukú and Pirahã languages have quantitative words but these words don't indicate stable numerical quantities. An English comparison might be the terms "couple" and "handful." Anthropological studies have argued that monolingual speakers of these languages draw on only the evolutionary elements of numerical thinking that are present in newborns and other members of the animal kingdom and thus perform approximate, rather than exact, quantity comparisons and arithmetical operations. However, when introduced to numerical symbols, members of these linguistic groups become capable of performing the same exact mathematical operations as people raised in cultures with numerical symbols. Everett, *Numbers and the Making of Us,* 9–10, 22–23, 117. Dehaene, *The Number Sense,* 263. Ian D. Holloway and Daniel Ansari, "Numerical Symbols: An Overview of Their Cognitive and Neural Underpinnings," in Roi Cohen Kadosh and Ann Dowker, eds., *The Oxford Handbook of Numerical Cognition* (Oxford: Oxford University Press, 2015), 539.

26. Dehaene, *The Number Sense,* 81, 264. Geoffrey B. Saxe, "Culture, Language, and Number," in Roi Cohen Kadosh and Ann Dowker, eds., *The Oxford Handbook of Numerical Cognition* (Oxford: Oxford University Press, 2015), 370. For example, the Aranda people in Australia have number words only for one, two, and three. For larger numbers they have a word meaning "this many, as many as indicated," which is accompanied by drawing the appropriate number of lines in the sand. Butterworth, *What Counts,* 50, 154–5.

27. Helen De Cruz and Johan De Smedt, "Mathematical Symbols as Epistemic Actions," *Synthese* 190, no. 1 (January 1, 2013): 4. Stanislas Dehaene and Laurent Cohen, "Cultural Recycling of Cortical Maps," *Neuron* 56, no. 2 (October 25, 2007): 384–98. Sami Abboud et al., "A Number-Form Area in the Blind," *Nature Communications* 6, no. 1 (January 23, 2015): 1–9. Marie Amalric, Isabelle Denghien, and Stanislas Dehaene, "On the Role of Visual Experience in Mathematical Development: Evidence from Blind Mathematicians," *Developmental Cognitive Neuroscience* 30 (April 1, 2018): 314–23. Marie Amalric and Stanislas Dehaene, "A Distinct Cortical Network for Mathematical Knowledge in the Human Brain," *NeuroImage* 189 (April 1, 2019): 19–31. Marie Amalric and Stanislas Dehaene, "Cortical Circuits for Mathematical Knowledge: Evidence for a Major Subdivision within the Brain's

Semantic Networks," *Philosophical Transactions of the Royal Society B: Biological Sciences* 373, no. 1740 (February 19, 2018). Marie Amalric and Stanislas Dehaene, "Origins of the Brain Networks for Advanced Mathematics in Expert Mathematicians," *Proceedings of the National Academy of Sciences* 113, no. 18 (May 3, 2016): 4909–17. This stands in contrast with literacy skills, which are mapped onto regions of the brain associated with vision, specifically object and scene recognition. When someone acquires the ability to read, the brain has to reorganize itself, "recycling" parts of itself to support this new skill. Stanislas Dehaene and Laurent Cohen, "Cultural Recycling of Cortical Maps," *Neuron* 56, no. 2 (October 25, 2007): 384–98. Stanislas Dehaene et al., "Illiterate to Literate: Behavioural and Cerebral Changes Induced by Reading Acquisition," *Nature Reviews Neuroscience* 16, no. 4 (April 2015): 234–44.

28. Bulwer, *Chirologia*, 188-9.
29. English does have words with both numerical and non-numerical meanings. For example, "gross" can mean the quantity one hundred and forty four—that is, twelve twelves, an important number in base twelve systems of arithmetic. The word "gross" also has a host of associated definitions, including large, easy to understand, and the entirety of some group. An even greater range of definitions can be found for "brace," which can describe a piece of armor, part of a bridge, or a type of fastening device but is also used to mean two of something. However, none of these words are in the main counting sequence; we say, "one, two, three," not "one, brace, three." *OED*, s.vv. "gross, a. and n."[4] "brace, n."[2]
30. Paul Sillitoe, ed., *Local Science vs. Global Science: Approaches to Indigenous Knowledge in International Development* (New York: Berghahn Books, 2007), 265. Calvin C. Clawson, *The Mathematical Traveler: Exploring the Grand History of Numbers* (New York: Plenum Press, 1994), 28.
31. While the steps that children go through appear to be stable across cultures and socioeconomic status, the timing is not. Twenty-first-century American children make the association between symbols and abstract quantities somewhere between the ages of three and six, usually around age four. Korean children learn two different sets of number words and are slower in making the initial connection between words and quantities, though by age six or seven they surpass the numerical abilities of their American peers. Smaller cultural variations within linguistic groups—such as those caused by socioeconomic status—can also cause differences in timing. Sarnecka, Goldman, and Slusser, "Children's First Representations," 297. Dehaene, *The Number Sense*, 106–7. Yukari Okamoto, "Mathematical Learning in the USA and Japan: Influences of Language," in Roi Cohen Kadosh and Ann Dowker, eds., *The Oxford Handbook of Numerical Cognition* (Oxford: Oxford University Press, 2015), 424. Dehaene, *The Number Sense*, 127. Geetha B. Ramani and Robert S. Siegler, "How Informal Learning Activities Can Promote Children's Numerical Knowledge," in Roi Cohen Kadosh and Ann Dowker, eds., *The Oxford Handbook of Numerical Cognition* (Oxford: Oxford University Press, 2015), 1142-1143.
32. Barbara W. Sarnecka, Meghan C. Goldman, and Emily B. Slusser, "How Counting Leads to Children's First Representations of Exact, Large Numbers," in Roi Cohen Kadosh and Ann Dowker, eds., *The Oxford Handbook of Numerical Cognition*

(Oxford: Oxford University Press, 2015), 292, 299. Everett, *Numbers and the Making of Us,* 160.

33. Other bases are possible, though systems intended for direct human use have histor-ically been some multiple of 10. Examples include the traditional base 20 systems of Mesoamerican and Iñupiaq languages, as well as Mesopotamians' base 60. Stephen Chrisomalis, *Numerical Notation: A Comparative History* (Cambridge: Cambridge University Press, 2010), 363, 377.

34. James R. Hurford, *Language and Number: The Emergence of a Cognitive System* (New York: Basil Blackwell, 1987).

35. English is not the only language with such irregularities. In Japanese, the number 25 maps regularly to a number word meaning two-tens-five (2*10 + 5), but in French the number is twenty-five (20 + 5) and in Dutch it is inverted as five-and-twenty (5 + 20). Japanese and Chinese children, whose number words map more perfectly to decimal numbers, gain an understanding of place value about two years earlier than American children using English number words which gives them a numerical ad-vantage that lasts for years. Modern Welsh number words also map more closely to the decimal system than English words, and comparative studies of Welsh- and English-speaking children from the same region have shown that the Welsh-speaking children have a similar early advantage. John N. Towse, Kevin Muldoon, and Victoria Simms, "Figuring Out Children's Number Representations: Lessons from Cross-Cultural Work," in Roi Cohen Kadosh and Ann Dowker, eds., *The Oxford Handbook of Numerical Cognition* (Oxford: Oxford University Press, 2015), 404. Okamoto, "Mathematical Learning," 419–21. Stephen Chrisomalis, *Reckonings: Numerals, Cognition, and History* (Cambridge, MA: MIT Press, 2020), 76–77.

36. "Twelve" is similarly irregular. The number 13 is inverted as "thirteen" or three-ten (3 + 10) instead of ten-three (10 + 3), and the other "teen" numbers are also inverted. While the decade numbers are more regular, once "ty" is substituted for "ten" in words like "sixty" and "eighty," there are still slight variations such as "two" becoming "twen" in "twenty."

37. Bulwer, *Chirologia,* 1845.

38. The counting numbers—also known as natural numbers—begin with 1, 2, 3 and con-tinue indefinitely through the positive integers (whole numbers greater than 1).

39. Hurford, *Language and Number.*

40. While Recorde stated he did not need to teach people 1 through 5 times 1 through 5, his multiplication tables do list all the multiples of 1 through 9 times 1 through 9. Recorde, *The Ground of Artes* (1543), C1v-C2r, D3r, G5v.

41. Neugebauer, "Mental Handicap," 29–30.

42. Butterworth, Varma, and Laurillard, "Dyscalculia: From Brain to Education," 655. Brian Butterworth, "The Implications for Education of an Innate Numerosity-Processing Mechanism," *Philosophical Transactions of the Royal Society B: Biological Sciences* 373, no. 1740 (February 19, 2018). Avishai Henik, Orly Rubinsten, and Sarit Ashkenazi, "Developmental Dyscalculia as a Heterogeneous Disability," in Roi Cohen Kadosh and Ann Dowker, eds., *The Oxford Handbook of Numerical Cognition* (Oxford: Oxford University Press, 2015), 665. Annemie Desoete, "Cognitive

Predictors of Mathematical Abilities and Disabilities," in Roi Cohen Kadosh and
Ann Dowker, eds., *The Oxford Handbook of Numerical Cognition* (Oxford: Oxford
University Press, 2015), 916.

43. Michèle M. M. Mazzocco, "The Contributions of Syndrome Research to the Study
of MLD," in Roi Cohen Kadosh and Ann Dowker, eds., *The Oxford Handbook of
Numerical Cognition* (Oxford: Oxford University Press, 2015), 681.

44. Lien Peters and Bert De Smedt, "Arithmetic in the Developing Brain: A Review
of Brain Imaging Studies," *Developmental Cognitive Neuroscience* 30 (April 1,
2018): 265–79.

Chapter 1

1. Keith Thomas, "Numeracy in Early Modern England: The Prothero Lecture,
Read 2 July 1986," in *Transactions of the Royal Historical Society*, 5th Series, no. 37
(London: Royal Historical Society, 1987), 120.

2. Robert Recorde, *The Ground of Artes Teachyng the Worke and Practise of Arithmetike*
(London: R. Wolfe, 1543), T2r. Over one hundred structurally distinct symbolic systems are known to have been used between 3500 BC and the present day. Many of these
systems coexisted in time and space; the historical peak of system diversity occurred
around the turn of the sixteenth century. Stephen Chrisomalis, *Numerical Notation: A
Comparative History* (Cambridge: Cambridge University Press, 2010), 425.

3. Recorde, *The Ground of Artes* (1543), Q2v.

4. Children as young as six can successfully map dot arrays onto Arabic numerals and
vice versa. Ian D. Holloway and Daniel Ansari, "Numerical Symbols: An Overview of
Their Cognitive and Neural Underpinnings," in Roi Cohen Kadosh and Ann Dowker,
eds., *The Oxford Handbook of Numerical Cognition* (Oxford: Oxford University Press,
2015), 538.

5. Recorde, *The Ground of Artes* (1543), A2r-v.

6. Thomas Hobbes, *Leviathan, or, The matter, forme, and power of a common wealth,
ecclesiasticall and civil* (London: Andrew Crooke, 1651), 14. Some linguists, such as
Heike Wiese, similarly argue that the concept of numbers evolved from human language: number words "do not refer to numbers, they *serve as* numbers." Heike Wiese,
Numbers, Language, and the Human Mind (Cambridge: Cambridge University Press,
2003), 5. However, neuroscientific studies seem to disprove this hypothesis.

7. John Brinsley, *Lvdvs Literarivs: or, the Grammar Schoole; shewing how to proceede
from the first entrance into learning, to the highest perfection required in the Grammar
Schooles* (London: Felix Kyngston, 1627), 26.

8. John Hall, *Of government and obedience as they stand directed and determined by
Scripture and reason* (London: T. Newcomb, 1654), 265–6.

9. David C. Geary, "The Classification and Cognitive Characteristics of Mathematical
Disabilities in Children," in Roi Cohen Kadosh and Ann Dowker, eds., *The Oxford
Handbook of Numerical Cognition* (Oxford: Oxford University Press, 2015), 773.

10. Brian Butterworth, *What Counts: How Every Brain Is Hardwired for Math* (New York: Free Press, 1999), 211. Michael Andres and Mauro Pesenti, "Finger-Based Representation of Mental Arithmetic," in Roi Cohen Kadosh and Ann Dowker, eds., *The Oxford Handbook of Numerical Cognition* (Oxford: Oxford University Press, 2015), 73.
11. Andres and Pesenti, "Finger-Based Representation of Mental Arithmetic," 72.
12. The twenty-eight years of the solar cycle were counted using the fourteen joints on each hand. The nineteen years of the lunar cycle were counted by using the fourteen joints and five fingertips on one hand. Georges Ifrah, *From One to Zero: A Universal History of Numbers*, trans. Lowell Bair (New York: Penguin Books, 1985), 62–3.Margaret E. Schotte, *Sailing School: Navigating Science and Skill, 1550–1800* (Baltimore: Johns Hopkins University Press, 2019), 35–7.
13. Stanislas Dehaene, *The Number Sense: How the Mind Creates Mathematics* (New York: Oxford University Press, 1997), 93. Ifrah, *From One to Zero*, 11.Caleb Everett, *Numbers and the Making of Us: Counting and the Course of Human Cultures* (Cambridge, MA: Harvard University Press, 2017), 81. Andres and Pesenti, "Finger-Based Representation of Mental Arithmetic," 70.
14. Modern scholars have similarly theorized that base 20 numerical systems derive from counting on fingers and toes. Dehaene, *The Number Sense*, 83.
15. Hobbes, *Leviathan*, 14. Similar theories had been advanced since the ancient Greeks; see, for example, Herodotus and Aristotle. John Fauvel and Jeremy Gray, eds., *The History of Mathematics: A Reader* (New York: Palgrave MacMillan, 1987), 2.
16. Stephen Chrisomalis has calculated that approximately 90% of human symbolic systems are base 10, with base 20 being the next most common at about 7%. There are five addition systems that have base 60 or 100. Chrisomalis, *Numerical Notation*, 379.
17. John Bulwer, *Chirologia, or, The naturall language of the hand composed of the speaking motions, and discoursing gestures thereof: whereunto is added Chironomia, or, The art of manuall rhetorick.* (London: Thomas Harper, 1644), 184. Modern studies have shown that people from English-speaking countries usually begin counting on the left hand, while people from Arabic-speaking countries in the Middle East usually begin with their right hand. The direction of counting may also be influenced by right- and left-handedness. Andres and Pesenti, "Finger-Based Representation of Mental Arithmetic," 76.
18. Bulwer, *Chirologia*, 185.
19. By contrast, in twenty-first-century America, the use of finger counting is positively correlated with accuracy in younger children but becomes negatively correlated with accuracy by about the end of second grade. If older children continue to use finger counting, it's considered a sign that they are having difficulty with mathematics. Jo-Anne Lefevre, Emma Wells, and Carla Sowinski, "Individual Differences in Basic Arithmetical Processes in Children and Adults," in Roi Cohen Kadosh and Ann Dowker, eds., *The Oxford Handbook of Numerical Cognition* (Oxford: Oxford University Press, 2015), 905–6. Despite people generally eschewing the use of finger counting in adulthood, brain imaging studies have demonstrated that neurological links remain between arithmetical tasks and finger movements. Andres and Pesenti, "Finger-Based Representation of Mental Arithmetic," 82–3.

20. Diego Hurtado de Mendoza, *The pleasant adventures of the witty Spaniard, Lazarillo de Tormes* (London: J. Leake, 1688), 90. Pierre de La Primaudaye, *The French Academie Fully discoursed and finished in foure bookes* (London: Thomas Adams, 1618), 203.

21. George Brown, *A Compendious, but a Compleat System Of Decimal Arithmetick, Containing more Exact Rules for ordering Infinities, than any hitherto extant* (Edinburgh: George Brown, 1701), 3–4.

22. Margaret Cavendish, *Natures picture drawn by fancies pencil to the life* (London: A. Maxwell, 1671), 166.

23. Mary Astell, *An Essay In Defence of the Female Sex. In which are inserted the CHARACTERS OF A Pedant, A Squire, A Beau, A Vertuoso, A Poetaster, A City-Critick, &c.* (London: A. Roper, E. Wilkinson and R. Clavel, 1696), 120.

24. Bulwer, *Chirologia*, 87, 85.

25. The first chapter of Bede's *De temporum ratione* was entitled *De computatio vel loquela digitorum* and was devoted to finger counting. Elisabeth Alföldi-Rosenbaum, "The Finger Calculus in Antiquity and in the Middle Ages: Studies on Roman Game Counters I," *Frühmittelalterliche Studien* 5, no. 1 (1971): 1–9. Burma P. Williams and Richard S. Williams, "Finger Numbers in the Greco-Roman World and the Early Middle Ages," *Isis* 86 (1995): 588–90, 592. Scott G. Bruce, *Silence and Sign Language in Medieval Monasticism: The Cluniac Tradition, c. 900–1200* (Cambridge: Cambridge University Press, 2010), 57–8.

26. Williams and Williams, "Finger Numbers," 592–4. Moritz Wedell, "Numbers," in Albrecht Classen, ed., *Handbook of Medieval Culture*, vol. 2 (Berlin: De Gruyter, 2015), 1237. Karl Menninger, *Number Words and Number Symbols: A Cultural History of Numbers*, trans. Paul Bronner (Cambridge, MA: MIT Press, 1969), 211. Stephen Chrisomalis, "The Cognitive and Culture Foundations of Numbers," in Eleanor Robson and Jacqueline Stedall, eds., *The Oxford Handbook of the History of Mathematics* (Oxford: Oxford University Press, 2009): 504–5.

27. Recorde, *The Ground of Artes* (1543), T2r.

28. Bulwer similarly created a system of alphabetical and numerical gestures that was "so ordered to serve for privy cyphers for any secret intimation," rather than designed to facilitate public communication in the same manner as other symbolic systems. Bulwer, *Chirologia*, 187.

29. W. T. Baxter, "Early Accounting: The Tally and the Checkerboard," *Accounting Historians Journal* 16, no. 2 (December 1989): 65. Thomas, "Numeracy," 119–20.

30. Chrisomalis, "Foundations of Numbers," 502–3.

31. The British Museum has several Roman tallies made of bronze and bone/ivory—see, for example, Item Registration numbers 1868,0520.39; 1873,0820.646; 1814,0704.1081; 1772,0311.8; 1814,0704.1080; 1859,0301.51—while we still tally by drawing lines on a piece of paper today. Menninger, *Number Words*, 224–31. Baxter, "Early Accounting," 43–83. Wedell, "Numbers," 1207.

32. A mestizo, Inca Garcilasso de la Vega, spent much of his life in Spain and is most famous for his chronicles of his mother's people. Inca Garcilasso de la Vega, *The Royal Commentaries of Peru in Two Parts* (London: Miles Flesher for Samuel Heyrick, 1688), 157. John Smith, *The Generall Historie of Virginia, New-England, and the Summer Isles with the names of the Adventurers, Planters, and Governours from their*

first beginning. Ano: 1584 to this present 1624. (London: I. D. and I. H. for Michael Sparkes, 1624), 143.

33. Baxter, "Early Accounting," 47. Samuel Purchas, *Purchas His Pilgrimes. In Five Bookes. The third, Voyages and Discoueries of the North parts of the World, by Land and Sea, in Asia, Evrope, the Polare Regions, and in the North-west of America* (London: William Stansby for Henri Fetherstone, 1625), 92. Lewes Roberts, *The Merchants Mappe of Commerce: Wherein The Vniuersall Manner and Matter of Trade, is compendiously handled* (London: R. O for Ralh Mabb, 1638), 174.

34. Caroline Shenton, *The Day Parliament Burned Down* (Oxford: Oxford University Press, 2012), 50. David A. Carpenter, *The Struggle for Mastery: Britain, 1066–1284* (London: Penguin Books, 2005), 154. Menninger, *Number Words,* 228.

35. The *Dialogus de Saccario* explains the proper "length of a tally is from the tip of the index finger to the tip of the thumb extended," yet some surviving examples held by the Bank of England are eight and a half feet long. Hilary Jenkinson, "Exchequer Tallies," in *Archaeologia or, Miscellaneous Tracts Relating to Antiquity,* 2nd ser., 62 (1911): 373. Hilary Jenkinson, "Medieval Tallies, Public and Private," in *Selected Writings of Sir Hilary Jenkinson* (Gloucester, England: Alan Sutton , 1980), 63. Exchequer tallies had four sides, but instances exist of eight or even sixteen-sided tallies. Baxter, "Early Accounting," 46.

36. Hilary Jenkinson, "An Original Exchequer Account of 1304 with private tallies attached," in *Proceedings of the Society of Antiquaries of London, 27th November 1913 to 25th June 1914, Second Series, Vol. XXVI* by Society of Antiquaries of London (Oxford: Horace Hart, 1914), 40. Baxter, "Early Accounting," 61.

37. Theophilus Field, *A Christians Preparation to the Worthy Receiuing of the Blessed Sacrament of the Lords Supper* (London: Augustine Mathewes for Tho. Thorp, 1622), 13. Thomas Scot, *Philomythie, or Philomythologie wherein Outlandish Birds, Beasts, and Fishes, are taught to Speake true English plainely* (London: Francis Constable, 1622), K8v. William Shiers, *A Familiar Discourse or Dialogue Concerning the Mine-Adventure* (London: 1700), 127. Adam Fox, *Oral and Literate Culture in England 1500–1700* (Oxford: Clarendon Press, 2000), 21.

38. Baxter, "Early Accounting," 62. Thomas, "Numeracy," 119. *OED,* s.v.v. "stock, *n.1* and *adj.*" and "stock-jobber, *n.*"

39. This was most likely to occur when private individuals had pre-existing experience with Exchequer tallies. Jenkinson, "Exchequer Tallies," 377–9.

40. Jenkinson, "Exchequer Tallies," 373.

41. William Shakespeare, *The First Part of the Contention Betwixt the Two Famous Houses of Yorke and Lancaster with the Death of the Good Duke Humphrey: and the Banishment and Death of the Duke of Suffolke, and the Tragical End of the Prowd Cardinall of Winchester, with the Notable Rebellion of Iacke Cade: and the Duke of Yorkes First Clayme to the Crowne* (London: Valentine Simmes for Thomas Millington, 1600), F4r. Roger North, *The Gentleman Accomptant: OR, AN ESSAY To unfold the Mystery of Accompts* (London: E. Curll, 1714), b4r.

42. Baxter, "Early Accounting," 49.

43. Oliver Cromwell, *An Ordinance For bringing the Publique Revenues Of this Commonwealth into one Treasvry* (London: William du-Gard and Henry Hills, 1654), 396.

44. Shenton, *Parliament Burned*, 52. Menninger, *Number Words*, 237. John Cowell, *The Interpreter: Or Booke Containing the Signification of Words* (Cambridge: Iohn Legate, 1607), Ssslr-v.

45. The original, Latin text reads: "Sussex. De Johanne Perpount et Johanne Yerman collectoribus custumarum et subsidorum domini Regis iń portu ville Cicestrie vij libras de eisdem custumis et subsidiis" and "pro domino de Bourchier per restitutionem vnius tallie videlicet xvij⁰ die Februarii anno xxiij⁰ Regis nunc leuate per manus Ricardi Wode." Jenkinson, "Exchequer Tallies," 370–1.

46. Hatfield House Archives, CP 109/79, https://search.proquest.com/docview/185 8039942.

47. For more on the scarcity of coinage in the early modern period and the use of credit as an alternative means of exchange, see Craig Muldrew, *The Economy of Obligation: The Culture of Credit and Social Relations in Early Modern England* (New York: St. Martin's Press, 1998), 98–101.

48. For more on the link between social trust and economic credit, see Muldrew, *The Economy of Obligation*, 148, 152–3. For more on calculating worth in early modern England, see Alexandra Shepard, *Accounting for Oneself: Worth, Status, and the Social Order in Early Modern England* (Oxford: Oxford University Press, 2015).

49. Hatfield House Archives, CP 30/80, https://search.proquest.com/docview/185 8026242.

50. Samuel Pepys, *The Diary of Samuel Pepys*, ed. Robert Latham and William Matthews (Berkeley: University of California Press, 1972), 6:70. For more on private banking, see Muldrew, *The Economy of Obligation*, 115.

51. Baxter, "Early Accounting," 54.

52. Many of Samuel Pepys' diary entries on tally sticks involved this sort of trading on future income. See, for example, Pepys, *Diary*, 6:133, 154, 157, 163–4; 7: 170, 407. Baxter, *Early Accounting*, 79. John Vernon, *The Compleat Comptinghouse: OR, The young Lad taken from the Writing School, and fully instructed, by way of Dialogue, in all the Mysteries of a Merchant* (London: J. D. for Benj. Billingsley, 1678), 124–6. Thomas Manby, *A collection of the statutes made in the reigns of King Charles the I. and King Charles the II. with the abridgment of such as stand repealed or expired* (London: John Streater, James Flesher, and Henry Twyford, 1667) 54. Fabian Philipps, *The Pretended Perspective-Glass* (London: n.p., 1669), 10–11. The fact that charging interest occurred at all was an innovation in the early modern period, and allowed only after 1571. Muldrew, *The Economy of Obligation*, 114.

53. William Williams, *Votes of the House of Commons, Perused and Signed to be Printed According to the Order of the House of Commons* (London: 1680), 165.

54. Michael Godfrey, *A Short Account of the Bank of England* (London: John Whitlock, 1695), 2. Jenkinson, "Exchequer Tallies," 371. R. D. Richards, *The Early History of Banking in England* (London: P. S. King & Son, 1929), 205–6, 230.

55. Richards, *Banking*, 154. Baxter, "Early Accounting," 79–80.

56. As early as the tenth century, medieval English officials had begun to produce "divided" or "toothed papers," in emulation of tally stick splitting conventions. Menninger, *Number Words*, 232.

57. William Cobbett, *The Parliamentary History of England, From the Earliest Period to the Year 1803*, vol 23 (London: R. Bagshaw, 1814), 120. Jenkinson, "Exchequer Tallies," 368–9. For a thorough examination of the Westminster tally stick fire, see Shenton, *Parliament Burned*.

58. The English name "jetton" comes from the French "jeton," which was derived from the verb "jeter," which means "to push," referencing the action of pushing counters during calculations. It fell out of use during the late fourteenth and fifteenth centuries as England lost its French territories. Francis Pierrepont Barnard, *The Casting-Counter and the Counting-Board: A Chapter in the History of Numismatics and Early Arithmetic* (Oxford: Clarendon Press, 1916), 26–7. J. M. Pullan, *The History of the Abacus* (London: Hutchinson, 1970), 73. *OED*, s.v.v. "counter, *n.3*" and "jetton, *n.*"

59. David Eugene Smith, *Computing Jetons* (New York: American Numismatic Society, 1921), 65. Barnard, *Casting-Counter*, 26. Chrisomalis, "Foundations of Numbers," 504.

60. Pullan, *Abacus*, 2, 16–17. Smith, *Computing Jetons*, 9. *OED*, s.v.v. "calculate, *v.1*" and "abacus, *n.*"

61. Pullan, *Abacus*, 68. Carpenter, *Struggle for Mastery*, 154.

62. Pullan, *Abacus*, 71-2. Barnard, *Casting-Counter*, 25. Smith, *Computing Jetons*, 28, 65–6, 68. M. B. Mitchiner, C. Mortimer, and A. M. Pollard, "Nuremberg and Its Jetons, c. 1475 to 1888: Chemical Compositions of the Alloys," *Numismatic Chronicle* 147 (1987): 115–16.

63. Pullan, *Abacus*, 51, 72, 75. Barnard, *Casting-Counter*, 63. Mitchiner, Mortimer, and Pollard, "Nuremberg and Its Jetons," 117–18. David Yoon, "Counting Tokens from the Excavations at Psalmodi (Gard, France)," *American Journal of Numismatics* 16/17 (2004–5), 173. Charles E. Orser, *An Archaeology of the English Atlantic World, 1600–1700* (Cambridge: Cambridge University Press, 2018), 207. Marjorie H. Akin, James C. Bard, and Kevin Akin, *Numismatic Archaeology of North America: A Field Guide to Historical Artifacts* (New York: Routledge, 2016), 84. J. C. Harrington, "Evidence of Manual Reckoning in The Cittie of Ralegh," *North Carolina Historical Review* 33 no. 1 (Jan.uary1956): 1.

64. Barnard, *Casting-Counter*, 32-3.

65. Barnard, *Casting-Counter*, 83. Pullan, *Abacus*, 73. Smith, *Computing Jetons*, 32. William K. Boyd and Henry W. Meikle, eds., *Calendar of State Papers relating to Scotland and Mary, Queen of Scots, 1547–1603*. Vol. 10: 1589–1593 (Edinburgh, Scotland: H. M. General Register House, 1936), 328, accessed via the State Papers Online, Gale Document Number MC4308800433. J. Gairdner, ed., *Letters and Papers, Foreign and Domestic, of the Reign of Henry VIII*. Vol. 13: Part II: 1538 (London: Her Majesty's Stationery Office, 1893), 499 or TNA SP 1/141 f.1, accessed via the State Papers Online, Gale Document Number MC4302301212. J. Gairdner, ed. *Letters and Papers, Foreign and Domestic, of the Reign of Henry VIII*. Vol. 9: 1535 (London: Longman, 1886), 99, accessed via the State Papers Online, Gale document number MC4301700304.

66. BL Cotton Cleopatra F/II. f.130, accessed via the State Papers Online, Gale Document Number MC4301700576. R.A. Roberts, ed. *Calendar of the Manuscripts of the Most*

Hon. the Marquis of Salisbury, Preserved at Hatfield House, Hertfordshire. Vol. 10:1600 (London: His Majesty's Stationery Office, 1904), 356, accessed via the State Papers Online, Gale Document Number MC4306100906.

67. Latten was composed of 60% copper, 30% zinc, and 10% lead. Barnard, *Casting-Counter*, 28, 91. For a detailed analysis of the chemical composition of counters, see Mitchiner, Mortimer, and Pollard, "Nuremberg and Its Jetons," 114–55 and M. B. Mitchiner, C. Mortimer, and A. M. Pollard, "The Alloys of Continental Copper-Base Jetons (Nuremberg and Medieval France Excepted)," *Numismatic Chronicle* 148 (1988):117–28. J. S. Brewer, ed., *Letters and Papers, Foreign and Domestic, of the Reign of Henry VIII.* Vol. 3: Part I: 1519–21 (London: Longman, Green, Reader and Dyer, 1867), 136 or TNA SP 1/18 f. 266, acessible via the State Papers Online, Gale Document Number MC4300700395.

68. In 1916, Francis Pierrepont Barnard's personal collection was approximately 7,000 counters and he surveyed over 30,000 more. Barnard, *Casting-Counter*, 7. Today, medieval and early modern counters are regularly available for purchase on websites like Ebay.

69. Recorde, *The Grounde of Artes* (1543), O2v.

70. *OED*, s.v. "counter, n.3." Randle Holme, *The Academy of Armory, or A Storehouse of Armory and Blazon* (Chester: Randle Holme, 1688), 259. Smith, *Computing Jetons*, 33.

71. Pullan, *Abacus*, 60.

72. Smith, *Computing Jetons*, 26.

73. Pullan, *Abacus*, 66, 69.

74. *OED*, s.v. "counting-house, n."

75. Barnard, *Casting-Counter*, 245.

76. Menninger, *Number Words*, 334, 378. Counters were a standard New Year's, birthday, or anniversary gift, especially in France and the Low Countries. Barnard, *Casting-Counter*, 6, 77, 79. HEH, MS HA Inventories Box 1, Folder 1, f. 13.

77. Recorde, *The Grounde of Artes* (1543), S7v-S8r.

78. There seems to have been some variation in whether base 20 or base 10 numbers applied to this and subsequently higher lines. Recorde, *The Grounde of Artes* (1543), S8v, T1v. Pullan, *Abacus*, 66.

79. For the slightly different layout of counters in an auditors' form account, see Figure 1.5.

80. Recorde, *The Grounde of Artes* (1543), T1r.

81. A similar image is reproduced in the 1558 edition of *The Ground of Artes* but without the writing on the wall under the shelf. A different image, in the 1615 edition, shows two men and a boy calculating with counters alone.

82. The Arabic numerals on the right side of the table are struck-through as part of division calculations.

83. The VDMIE is most likely an abbreviation for *Verbum Domini Manet in Eternum*—the Word of the Lord shall remain forever. This was a slogan used by German Reformers in the 1520s. John M. Todd, *Luther: A Life* (London: Hamish Hamilton, 1982), 282.

84. Vernon, *Compleat Comptinghouse,* 16.

85. Barnard, *Casting-Counter*, 34.

86. Edward Hawkins, compiler, Augustus W. Franks and Herbert A. Brueber, eds., *Medallic Illustrations of the History of Great Britain and Ireland to the Death of George II* (London: Spink & Son, 1969), 1:7, 85. Barnard, *Casting-Counter*, 8, 47. The earliest of these Rechenmeister counters appears to date to around 1550. Pullan, *Abacus*, 81. The legend on the author's counter was translated by Nancy O. Chamness.

87. Hawkins, *Medallic Illustrations*, 1:55, 85.

88. Counters often served a secondary purpose as instruments of gaming. Elizabethan sources talk of children playing with points, pins, cherry-stones, and counters, all of which served the dual function of game tokens and a currency substitute to determine winners and losers. Nicholas Orme, *Medieval Children* (New Haven, CT: Yale University Press, 2001), 177-8. Hawkins, *Medallic Illustrations*, 1:375-6.

89. For numerous examples of personalized English counters, see Hawkins, *Medallic Illustrations*, 1:85, 101, 133, 146-7, 151-3, 160-2, 188-91, 196-7, 279, 286.

90. Pullan, *Abacus*, 76. Hawkins, *Medallic Illustrations*, 2:413-15, 464, 483.

91. Mitchiner, Mortimer, and Pollard, "Nuremberg and Its Jetons," 122. Portugal seems to have similarly abandoned counters around 1640. Francies Pierrepont Barnard, "Portuguese Jettons," *The Numismatic Chronicle and Journal of the Royal Numismatic Society* 5th Ser., 3 (1923): 76.

92. Pullan, *Abacus*, 55-6.

93. Barnard, *Casting-Counter*, 304. Pullan, *Abacus*, 1-2. Smith, *Computing Jetons*, 29.

94. Daines Barrington, "An Historical Disquisition on the Game of Chess; addressed to Count de Bruhl, F.A.S.," *Archaeologia* 9 (London: J. Nichols, 1789), 28 note g, https://doi.org/10.1017/S0261340900024474. Barnard, *Casting-Counter*, 252.

95. Pullan, *Abacus*, 83. Barnard, *Casting-Counter*, 85.

96. There were medieval variations on the system, meant to shorten the numerals, but they were not commonly used in the early modern period. For more on medieval number systems, see J. Lennart Berggren, "Medieval Arithmetic: Arabic Texts and European Motivations," in John J. Contreni and Santa Casciani, *Word, Image, Number: Communication in the Middle Ages* (Tavarnuzze (Florence), Italy: SISMEL Edizioni del Galluzzo, 2002), 352-3.

97. Roman numerals remain in general use to the present day in a limited number of contexts, such as numbering monarchs, numbering preface pages in a book, or numbering endnotes in a Microsoft Word document.

98. Stephen Chrisomalis, "The Origins and Co-Evolution of Literacy and Numeracy," in David R. Olson and Nancy Torrance, eds., *The Cambridge Handbook of Literacy* (Cambridge: Cambridge University Press, 2009), 60. Chrisomalis, *Numerical Notation*, 132.

99. Brinsley, *Lvdvs Literarivs*, 25-6.

100. The legacy of these assumptions can still be seen today. A 2011 popular history book commented on "the awkwardness of Roman numerals when performing arithmetic operations such as multiplication" while the Historic Jamestown website in 2020 described late sixteenth- and early seventeenth-century counters as "eas[ing] the complicated maneuvers needed for multiplying and dividing Roman numerals, but it was still a tedious process." Jim Al-Khalili, *The House of Wisdom: How Arabic Science Saved Ancient Knowledge and Gave Us the Renaissance* (New York: Penguin

Press, 2011), 1001. Historic Jamestown, "Jettons," https://historicjamestowne.org/collections/artifacts/jettons/, accessed June 15, 2020.

101. Prior to the widespread adoption of Arabic numeral arithmetic, the counting board was the dominant system for calculation in England. Thomas, "Numeracy," 120. John Denniss, "Learning Arithmetic: Textbooks and Their Users in England 1500–1900," in *The Oxford Handbook of the History of Mathematics*, ed. Eleanor Robson and Jacqueline Stedall (Oxford, 2009), 448–67.

102. Kim Plofker, *Mathematics in India* (Princeton, NJ: Princeton University Press, 2009), 255–6. Smith, *Computing Jetons*, 6. Wedell, "Numbers," 1212–3. The numerals for 2, 4, and 5 underwent particularly significant transformations over time. For images of the numbers' changing appearance, see G. F. Hill, *The Development of Arabic Numerals in Europe: Exhibited in 64 Tables* (Oxford: Clarendon Press, 1915) and David Eugene Smith and Louis Charles Karpinski, *The Hindu-Arabic Numerals* (Boston: Ginn and Company, 1911), 49, 69, 88, 140–8.

103. Wedell, "Numbers," 1213–4, 1238.

104. Peter Wardley and Pauline White, "The Arithmeticke Project: A Collaborative Research Study of the Diffusion of Hindu-Arabic Numerals," *Family and Community History* 6 (May 2003), 7. Chrisomalis, *Numerical Notation*, 223.

105. Chrisomalis, *Numerical Notation*, 120.

106. Hill, *Development*, 34–5, 54–61. Wedell, "Numbers," 1215.

107. Other scholars have also observed the early association of Arabic numerals with dates, e.g., Pullan, *Abacus*, 41.

108. Hill, *Development*, 61.

109. Barnard, *Casting-Counter*, 53–4. Hill, *Development*, 34–5 and 54–61.

110. C. C. Webb, ed., *The Churchwardens' accounts of St Michael, Spurriergate, York, 1518-1548* ([York]: University of York, Borthwick Institute of Historical Research, 1997), 2:327–8. William Mellows, ed., *Peterborough local administration; parochial government before the reformation. Churchwardens' accounts, 1467-1573, with supplementary documents, 1107-1488*, Publications of the Northamptonshire Record Society 9 (Kettering: Northamptonshire Record Society, 1939), 164.

111. James Croston, ed., *The register book of christenings, weddings, and burials, within the parish of Prestbury, in the county of Chester, 1560-1636*, ([Manchester]: Record Society of Lancashire and Cheshire, 1881), 188.

112. Esther M. E. Ramsay, ed., *The Churchwardens' Accounts of Walton-on-the-Hill, Lancashire 1627-1667* ([Liverpool]: Record Society of Lancashire and Cheshire, 2005), 4.

113. Muriel St. Clare Byrne, ed., *The Lisle Letters* (Chicago: University of Chicago Press, 1981), 4:191, 298, 514–15. Savage and Fripp, *Stratford-upon-Avon*, 2:2-39.

114. William Mellows, ed., *Peterborough Local Administration; Parochial Government from the Reformation to the Revolution, 1541-1689. Minutes and Accounts of the Feoffees and Governors of the City Lands with Supplementary Documents*, Publications of the Northamptonshire Record Society 10 (Kettering: Northamptonshire Record Society, 1937), 1–82.

115. Stephen G. Doree, ed., *The Parish Register and Tithing Book of Thomas Hassall of Amwell* ([Ware, Hertfordshire]: Hertfordshire Record Society, 1989), 81.

116. R.W. Greaves, ed., *The First Ledger Book of High Wycombe*, Buckinghamshire Record Society Publications, 11 (Hertfordshire: Broadwater Press, 1947), 110. Ramsay, *Walton-on-the-Hill*, 4.

117. Clive Burgess, ed., *The Church Records of St Andrew Hubbard, Eastcheap, c1450–c1570*, ([London]: London Record Society, 1999), 266.

118. F. Bailey, ed., *The Churchwardens' Accounts of Prescot, Lancashire, 1523–1607* (Preston: Record Society of Lancashire and Cheshire, 1953), 133. Mellows, *Feoffees*, 111.

119. F.W. Weaver and G.N. Clark, eds., *Churchwardens' Accounts of Marston, Spelsbury, Pyrton*, Oxfordshire Records Series, 6 (Oxford: Oxford University Press, 1925), 38.

 Robert Steele, *The Earliest Arithmetics in English* (London: Oxford University Press, 1922), xvii. Anthony Palmer, ed., *Tudor Churchwardens' Accounts*, Hertfordshire Record Society, 1 ([Ware, Hertfordshire]: Hertfordshire Record Society, 1985), 41, 177.

120. Chrisomalis, *Numerical Notation*, 122.

121. Wardley and White, "The Arithmeticke Project," 8, 12–13. More extensive data and reports from the Arithmeticke Project are available at https://web.archive.org/web/20090924015208/http://www.rw007a7896.pwp.blueyonder.co.uk:80/.

122. Wardley and White, "The Arithmeticke Project," 6. Pullan, *Abacus*, 39.

123. Wardley and White, "The Arithmeticke Project," 9–10.

124. *OED*, s.v.v. "algorism, *n.*" "cipher, *v.*"

125. John Palsgrave, *Lesclarcissement de la Langue Francoyse* (London: Richard Pynson and Iohan Haukyns, 1530), 188r.

126. Recent objects used to assist in calculations, which readers may have used at some point in their lives, include slide rules, mechanical and electronic calculators, and cellphones with calculator applications.

127. For more on eighteenth-century popular numeracy, see Benjamin Wardhaugh, *Poor Robin's Prophecies: A Curious Almanac, and the Everyday Mathematics of Georgian Britain* (Oxford: Oxford University Press, 2012).

Chapter 2

1. Gervase Markham, *The English Husbandman, drawn into two Bookes, and each Booke into two Parts* (London: William Sheares, 1635), 8–9.

2. Markham, *The English Husbandman*, 8–9.

3. *OED*, s.v. "cunning, *adj.*"

4. There are several early modern historians whose epistemological work on truth and facts has led them to examine the role of trust in the creation of knowledge. See, for example, Steven Shapin, *A Social History of Truth: Civility and Science in Seventeenth-Century England* (Chicago: University of Chicago Press, 1995); Barbara Shapiro, *A Culture of Fact: England, 1550–1720* (Ithaca, NY: Cornell University Press, 2000); Barbara Shapiro, *Probability and Certainty in Seventeenth-Century England: A Study of the Relationships Between Natural Science, Religion, History, Law, and Literature*

(Princeton, NJ: Princeton University Press, 1983). Mary Poovey also examined the intersection of trust in merchants and trust in the seemingly transparent double-entry bookkeeping system, while Craig Muldrew looked specifically about trust and economic credit. Mary Poovey, *A History of the Modern Fact: Problems of Knowledge in the Sciences of Wealth and Society* (Chicago: University of Chicago Press, 1998), 29–91. Craig Muldrew, *The Economy of Obligation: The Culture of Credit and Social Relations in Early Modern England* (New York: St. Martin's Press, 1998), 123, 129, 148–53.

5. Markham, *The English Husbandman*, 9.
6. Eugene Smith and Louis Charles Karpinski, *The Hindu-Arabic Numerals* (Boston: Ginn, 1911), 132.
7. Karl Menninger, *Number Words and Number Symbols: A Cultural History of Numbers*, trans. Paul Bronner (Cambridge, MA: MIT Press, 1969), 422.
8. Georges Ifrah, *From One to Zero: A Universal History of Numbers*, trans. Lowell Bair (New York: Viking Penguin, 1985), 481.
9. Menninger, *Number Words*, 422.
10. Muldrew, *The Economy of Obligation*, 98–101, 107–10.
11. Walter de Henley, *The Booke of Thrift, Containing a Perfite Order, and Right Methode to Profite Lands, And Other Things Belonging To Husbandry; Newly Englished, and Set Out By I.B. Gentleman of Caen in France* (London: Iohn Wolfe, 1589), B2r.
12. Theophilus Field, *A Christians Preparation To the Worthy Receiuing of the Blessed Sacrament of the Lord Supper* (London: Augustine Mathewes for Thomas Thorp, 1622), 13.
13. W. T. Baxter, "Early Accounting: The Tally and the Checkerboard," *Accounting Historians Journal* 16, no. 2 (December 1989): 62.
14. Thomas Scot, for example, complains of people lying about their tallies in the aftermath of fires that destroyed commercial buildings. Thomas Scot, *Philomythie, or Philomythologie wherin outlandish birds, beasts, and fishes, are taught to speake true English plainely* (London: Francis Constable, 1622), K8v.
15. Edward Chamberlayne, *The Second Part of the Present State of England Together With Divers Reflections Upon the Antient State Thereof* (London: Printed by T. N. for J. Martyn, 1671), 124, 128.
16. Hilary Jenkinson, "Exchequer Tallies," *Archaeologia or, Miscellaneous tracts relating to antiquity*, 2nd ser., 62 (1911): 374.
17. Thomas De Laune, *The Present State of London: or, Memorials Comprehending A Full and Succinct Account of the Ancient and Modern State Thereof* (London: George Larkin for Enoch Prosser and John How, 1681), 138.
18. Samuel Pepys, *The Diary of Samuel Pepys*, ed. Robert Latham and William Matthews (Berkeley: University of California Press, 1970–1983), 6:70.
19. Sarah Elestone, *The Last Speech and Confession of Sarah Elestone at the Place of Execution Who Was Burned for Killing Her Husband, April 24. 1678* (London: T. D., 1678), 2–3.
20. Thomas Wentworth, *The Office and Dutie of Executors* (London: T. C. for Andrew Crooke, Laurence Chapman, William Cooke, and Richard Best, 1641), 147–8. For more on the rise of the paper state, see Chapter 7.

21. William Sheppard, *The Touch-Stone of Common Assurances* (London: M. F. for W. Lee, M. Walbancke, D. Pakeman, and G. Bedell, 1648), 367. John Cowell, *The Institutes of the Lawes of England Digested Into the Method of the Civill or Imperiall Institutions* (London: Tho. Roycroft for Jo. Ridley, 1651), 181.

22. William Shiers, *A Familiar Discourse Or Dialogue Concerning the Mine-Adventure* (London: n.p., 1700), 127.

23. Pepys, *Diary*, 7:272.

24. Hugh Chamberlen, *Papers Relating to a Bank of Credit upon Land Security Proposed to the Parliament of Scotland* (Edinburgh: n.p., 1693), 4.

25. The use of IIII as opposed to IV was particularly common on dials. Jane Desborough, *The Changing Face of Early Modern Time, 1550–1770* (Cham, Switzerland: Palgrave Mcmillan, 2019), 63.

26. Stephen Chrisomalis, *Numerical Notation: A Comparative History* (Cambridge: Cambridge University Press, 2010), 119.

27. For more examples of how a modern reader might mistake one Arabic numeral for another, see Samuel A. Tannenbaum, *The Handwriting of the Renaissance: Being the Development and Characteristics of the Script of Shakespeare's Time* (New York: Frederick Ungar, 1967), 155–8.

28. Esther Ramsay, *John Isham, mercer and merchant adventurer: two account books of a London merchant in the reign of Elizabeth I* (Gateshead, Co. Durham: Northamptonshire Record Society, 1962). William Mellows, ed., *Peterborough local administration: parochial government from the Reformation to the Revolution, 1541–1689. Minutes and accounts of the feoffees and governors of the city lands with supplementary documents.* Publications of the Northamptonshire Record Society 10 (Kettering: Northamptonshire Record Society, 1937), 78.

29. J. M. Pullan, *The History of the Abacus* (London: Hutchinson, 1970), 34. Keith Thomas, "Numeracy in Early Modern England: The Prothero Lecture, Read 2 July 1986," in *Transactions of the Royal Historical Society*, 5th Series, no. 37 (London: Royal Historical Society, 1987), 121.

30. Techniques for performing calculations via Roman numerals have been invented but were never widely used. Chrisomalis, *Numerical Notation*, 115–6.

31. Stephen Chrisomalis, "The Cognitive and Culture Foundations of Numbers," in Eleanor Robson and Jacqueline Stedall, eds., *The Oxford Handbook of the History of Mathematics* (Oxford: Oxford University Press, 2009), 505. Patricia Cline Cohen, *A Calculating People: The Spread of Numeracy in Early America* (Chicago: University of Chicago Press, 1982), 27.

32. Early modern authors largely ignored the possibility that someone might have been born with fewer or more than ten fingers, or lost one or more fingers in accidents. It is likely that people developed idiosyncratic methods to calculate with nine or eleven fingers. Thomas Randolph, in his poem *Vpon the lose of his little finger*, compared nine fingers to the Arabic numerals 1 through 9 and joked the lost finger "did only for a Cipher stand." Thomas Randolph, *Poems with the Muses looking-glasse* (Oxford: Leonard Lichfield, 1638), 41.

33. For simple tally sticks, see, for example, several Roman tally sticks in the British Museum, registration numbers: 1868,0520.39; 1859,0301.51; 1856,1226.1494; 1772,0311.9; 1814,0704.1081; 1814,0704.1082; etc.

34. Modern scientists who study numerical cognition have shown that Arabic-numeral calculators use conceptual metaphors and fictive motion to help make the transition from understanding quantities to performing arithmetic, but in the case of counters, the numerical symbols actually physically moved. Caleb Everett, *Numbers and the Making of Us: Counting and the Course of Human Cultures* (Cambridge, MA: Harvard University Press, 2017), 202–6.

35. Such scholars do not appear to have considered the impact of European colonialism and cultural imperialism.

36. Ifrah, *From One to Zero*, 433.

37. Paul Lockhart, *Arithmetic* (Cambridge, MA: Harvard University Press, 2017), 61–2.

38. Helen De Cruz and Johan De Smedt, "Mathematical Symbols as Epistemic Actions," *Synthese* 190, no. 1 (January 1, 2013): 5. Similarly, animals and human infants can point to empty locations to indicate missing/absent objects. Joseph Tzelgov, Dana Ganor-Stern, Aravay Y. Kallai, and Michal Pinhas, "Primitives and Non-primitives of Numerical Representations," in Roi Cohen Kadosh and Ann Dowker, eds., *The Oxford Handbook of Numerical Cognition* (Oxford: Oxford University Press, 2015), 58.

39. While the counting-board form of the abacus is slower than Arabic numerals, not all abacus-style systems share this trait. A 1946 competition demonstrated that a Japanese soroban user was actually faster than an American using a state-of-the-art electronic calculator. Chrisomalis, "The Cognitive and Culture Foundations of Numbers," 506.

40. Francis Pierrepont Barnard, *The Casting-Counter and the Counting-Board: A Chapter in the History of Numismatics and Early Arithmetic* (Oxford: Clarendon Press, 1916), 304.

41. John Palsgrave, *Lesclarcissement de la Langue Francoyse* (London: Richard Pynson and Iohan Haukyns, 1530), 188r.

42. Calvin C. Clawson, *The Mathematical Traveler: Exploring the Grand History of Numbers* (New York: Plenum Press, 1994), 130; Menninger, *Number Words*, 224.

43. Peter Stallybrass, Roger Chartier, John Franklin Mowery, and Heather Wolfe, "Hamlet's Tables and the Technologies of Writing in Renaissance England," *Shakespeare Quarterly* 55, no. 4 (2004): 401, 405.

44. Robert Tittler, ed., *Accounts of the Roberts Family of Boarzell, Sussex: c1568–1582*, Sussex Records Series 71 (Lewes, England: Sussex Record Society, 1977), 59. In the original document, it is labeled folio 26v.

45. Anthony Palmer, ed., *Tudor Churchwardens' Accounts*, Hertfordshire Record Society, 1 (Ware, Hertfordshire: Hertfordshire Record Society, 1985), 16, 25–6, 79–84. F. A. Bailey, ed., *The Churchwardens' Accounts of Prescot, Lancashire 1523–1607* (Preston: Record Society of Lancashire and Cheshire, 1953), 135. Ramsay, *Isham*, cxi.

46. The mixing and matching of seventeenth-century systems appears to have slightly baffled Wardley and White, who note the "significant tension" between recording and calculation, while at the same time apologizing for the scribe in their earliest

example of a class three inventory as follows: "his command of the new system is almost complete, though Roman numerals creep in to the quantity descriptions." The idea that using Roman numerals might be a choice, rather than a deficit in learning, is mentioned later, but only briefly. Peter Wardley and Pauline White, "The Arithmeticke Project: A Collaborative Research Study of the Diffusion of Hindu-Arabic Numerals," *Family and Community History* 6 (May, 2003): 10, 13.

47. Wardley and White, "The Arithmeticke Project," 13.

48. Robert Recorde, *The Ground of Artes Teachyng the Worke and Practise of Arithmetike* (London: R. Wolfe, 1543), M6v. It is unclear if tables refer to erasable writing tables, such as wax tablets, or multiplication tables, such as the table provided in Recorde's *The Ground of Artes* (1543), F7r. Recorde did not expect his readers to have already memorized their one through ten times tables at the beginning of their course of study.

49. Similar dot notation appears in early modern Exchequer account books, likely reflecting the use of counters on the Exchequer "Table." Pullan, *Abacus*, 43–5.

Chapter 3

1. John Fauvel and Jeremy Gray, eds., *The History of Mathematics: A Reader* (New York: Palgrave Macmillan, 1987), 316–17. *Oxford Dictionary of National Biography*, s.v. "Wallis, John (1616–1703), *Mathematician and Cryptographer*," by Domenico Bertoloni Meli, http://www.oxforddnb.com/view/article/28572. The Wallis family's choice to send the elder son to a university and to set up younger sons to earn their livings through trade was common among early modern gentry. Patrick Wallis and Cliff Webb, "The Education and Training of Gentry Sons in Early Modern England," *Social History* 36, no. 1 (February 2011): 20–46.

2. For an analysis of the hyperbole in Wallis's autobiography, see Mordechai Feingold, *The Mathematicians' Apprenticeship: Science, Universities and Society in England, 1560–1640* (Cambridge: Cambridge University Press, 1984), 86–8.

3. Warren Van Egmond, *Practical Mathematics in the Italian Renaissance: A Catalog of Italian Abbacus Manuscripts and Printed Books to 1600* (Firenze: Instituto e Museo di Storia della Scienza, 1980), 9–12. The canonical work on late medieval commercial expansion is Robert S. Lopez, *The Commercial Revolution of the Middle Ages, 950–1350* (Englewood Cliffs, NJ: Prentice-Hall, 1971). For more recent sources, see, for example, David Ormrod, *The Rise of Commercial Empires: England and the Netherlands in the Age of Mercantilism, 1650–1770* (Cambridge: Cambridge University Press, 2003); and Sheilagh Ogilvie, *Institutions and European Trade: Merchant Guilds, 1000–1800* (Cambridge: Cambridge University Press, 2011). For more on late medieval Italian accounting practices, see R. H. Parker and B. S. Yamey's collection of post–Second World War British essays, especially G. A. Lee, "The Oldest European Account Book: A Florentine Bank Ledger of 1211"; C. W. Nobes, "The Gallerani Account Book of 1305–8"; and B. S. Yamey, "Balancing and Closing the Ledger: Italian Practice, 1300–1600," all in R. H. Parker and B. S. Yamey , eds., *Accounting History: Some*

British Contributions (Oxford: Oxford University Press, 1994), 160–267. A reprint of W. T. Baxter, "Early Accounting: The Tally and the Checker-board" *Accounting Historians Journal* 16, no. 2 (December 1989) can also be found in this volume. On accounting history more generally, see Ananias Charles Littleton and Basil S. Yamey, *Studies in the History of Accounting* (New York: Arno Press, 1978); and David Oldroyd and Alisdair Dobie, "Bookkeeping," in John Richard Edwards and Stephen P. Walker, eds., *The Routledge Companion to Accounting History* (New York: Routledge, 2009), 95–119.

4. Deborah E. Harkness, *The Jewel House: Elizabethan London and the Scientific Revolution* (New Haven, CT: Yale University Press, 2007), 101. Margaret E. Schotte, *Sailing School: Navigating Science and Skill, 1550–1800* (Baltimore: Johns Hopkins University Press, 2019), 11. For more on print and the dissemination of knowledge in England, see, for example, Elizabeth Eisenstein, *The Printing Press as an Agent of Change* (Cambridge: Cambridge University Press, 1979); Adrian Johns, *The Nature of the Book: Print and Knowledge in the Making* (Chicago: University of Chicago Press, 1998); and Jason Peacey, *Print and Public Politics in the English Revolution* (Cambridge: Cambridge University Press, 2013).

5. John Denniss, "Learning Arithmetic: Textbooks and Their Users in England 1500–1900," in Eleanor Robson and Jacqueline Stedall, eds., *The Oxford Handbook of the History of Mathematics* (Oxford: Oxford University Press, 2009), 448–67. For an example of a multiplication table included in a book with wax tablets, see Peter Stallybrass et al., "Hamlet's Tables and the Technologies of Writing in Renaissance England," *Shakespeare Quarterly* 55, no. 4 (Winter 2004): 397.

6. Islamic and Italian mathematicians used the dust board for Arabic-numeral arithmetic until pen and paper began to supplant it in the fourteenth century. Warren Van Egmond, "The Commercial Revolution and the Beginnings of Western Mathematics" (PhD diss., Indiana University, 1976), 343. Wax tablets continued to be produced and used for a variety of purposes, including calculation, throughout the early modern period. Stallybrass et al., "Hamlet's Tables," 402–3. While women did not perform calculations with their embroidery, they did reproduce letters and numbers on their samplers, and many surviving instances of women's handwriting show the clear influence of embroidered letters. Eleanor Hubbard, "Reading, Writing, and Initialing: Female Literacy in Early Modern London," *Journal of British Studies* 54, no. 3 (July 2015): 565.

7. Karl Menninger, *Number Words and Number Symbols: A Cultural History of Numbers*, trans. Paul Bronner (Cambridge, MA: MIT Press, 1969), 422; Georges Ifrah, *From One to Zero: A Universal History of Numbers*, trans. Lowell Bair (New York: Viking Penguin Inc., 1985), 481; Eugene Smith and Louis Charles Karpinski, *The Hindu-Arabic Numerals* (Boston: Ginn, 1911), 132.

8. David Cressy, *Literacy and the Social Order: Reading and Writing in Tudor and Stuart England* (Cambridge: Cambridge University Press, 1980), 142–56. John Arbuthnot. *An essay on the usefulness of mathematical learning, in a letter from a gentleman in the city to his friend in Oxford.* (Oxford: Anth. Peisley, 1701), 27.

9. Jonathan Barry, "Popular Culture in Seventeenth-Century Bristol," in *Popular Culture in Seventeenth-Century England*, ed. Barry Reay (London: Croom Helm, 1985), 62. Adam Fox, *Oral and Literate Culture in England 1500–1700* (Oxford: Clarendon Press, 2000), 18.

10. Heidi Brayman Hackel, "Popular Literacy and Society," in Joad Raymond, ed., *The Oxford History of Popular Print Culture*, vol. 1, *Cheap Print in Britain and Ireland to 1660* (Oxford: Oxford University Press, 2011), 93.

11. For more on the multiplicity of early modern literacies, see Keith Thomas, "The Meaning of Literacy in Early Modern England," in Gerd Baumann, ed., *The Written Word: Literacy in Transition, Wolfson College Lectures 1985* (Oxford: Clarendon Press, 1986), 99.

12. Cressy, *Literacy*, 142–56.

13. Hubbard, "Reading, Writing, and Initialing," 568–9.

14. Van Egmond, *Practical Mathematics*, 6–7.

15. Van Egmond, "Commercial Revolution," 72, 320–2, 341, 596. John V. Tucker, "Data, Computation and the Tudor Knowledge Economy," in Gareth Roberts and Fenny Smith, eds., *Robert Recorde: The Life and Times of a Tudor Mathematician* (Cardiff: University of Wales Press, 2012), 171-2.

16. Van Egmond, *Practical Mathematics*, 30–1.

17. Frank J. Swetz, *Capitalism and Arithmetic: The New Math of the 15th Century Including the Full Text of the Treviso Arithmetic of 1478*, trans. David Eugene Smith (La Salle, IL: Open Court, 1987), 24.

18. Van Egmond, "Commercial Revolution," 105–6; Ad Meskens, "Mathematics Education in Late Sixteenth-century Antwerp," *Annals of Science* 53, no. 2 (1996): 140.

19. John Barnard and Maureen Bell, "Appendix 1: Statistical Tables," in John Barnard and D. F. McKenzie, eds., *The Cambridge History of the Book in Britain*, vol. 4, *1557–1695* (Cambridge: Cambridge University Press, 2002), 779–85.

20. D. F. McKenzie, "Printing and Publishing, 1557–1700: Constraints on the London Book Trade," in John Barnard and D. F. McKenzie, eds., *Cambridge History of the Book in Britain*, vol. 4, *1557–1695* (Cambridge: Cambridge University Press, 2002), 556–8.

21. R. C. Simmons, "ABCs, Almanacs, Ballads, Chapbooks, Popular Piety and Textbooks," in John Barnard and D. F. McKenzie, eds., *Cambridge History of the Book in Britain*, vol. 4, *1557–1695* (Cambridge: Cambridge University Press, 2002), 504. For more on early modern almanacs, see Bernard Capp, *Astrology and the Popular Press: English Almanacs, 1500–1800* (Ithaca, NY: Cornell University Press, 1979); Timothy Feist, *The Stationers' Voice: The English Almanac Trade in the Early Eighteenth Century* (Philadelphia: American Philosophical Society, 2005); Alison A. Chapman, "Marking Time: Astrology, Almanacs, and English Protestantism," *Renaissance Quarterly* 60, no. 4 (Winter 2007): 1269–70; and Louise Hill Curth, *English Almanacs, Astrology and Popular Medicine, 1550–1700* (Manchester: Manchester University Press, 2007).

22. Fox, *Oral and Literate Culture*, 16. Margaret Spufford, *Small Books and Pleasant Histories: Popular Fiction and Its Readership in Seventeenth Century England* (London: Methuen, 1981), 101.

23. Thomas, "Meaning of Literacy," 112. Peacey, *Print and Public Politics*, 59. Fox, *Oral and Literate Culture*, 15.

24. Tessa Watt, *Cheap Print and Popular Piety, 1550–1640* (Cambridge: Cambridge University Press, 1991), 6. Spufford, *Small Books*, 66.

25. *English Short Title Catalog (ESTC)*, http://estc.bl.uk. Exact numbers are not possible to calculate due to the loss of sources over time and occasional difficulties distinguishing between editions printed in the same year. At least an additional seventy editions of basic accounting books were also produced during this period. B. S. Yamey, H. C. Edey, and Hugh W. Thomson, *Accounting in England and Scotland, 1543–1800: Double Entry in Exposition and Practice* (London: Sweet and Maxwell, 1963), 202–8. For more on mathematical publishing in general, see Harkness, *Jewel House*, 104. For the canonical work on early modern English mathematicians and their publications, see E. G. R. Taylor, *The Mathematical Practitioners of Tudor and Stuart England* (Cambridge: Cambridge University Press, 1954).

26. Harkness, *Jewel House*, 104–5. Meskens, "Mathematics Education," 152. Simon Schaffer, "Science," in Joad Raymond, ed., *Cheap Print in Britain and Ireland to 1660*, vol. 1 (Oxford: Oxford University Press, 2011), 399.

27. Edward Worsop, *A Discouerie of sundrie errours and faults daily committed by Lande-Meaters, ignorant of Arithmetike and Geometrie* (London: Henrie Middleton for Gregorie Seton, 1582), A2v. While Keith Thomas's reference to this passage carries with it a strong implication that arithmetic is too difficult to learn, the context of the full list makes it clear that Worsop was referring to Recorde's far less heralded geometry textbook, *The Pathway to Knowledge*. Keith Thomas, "Numeracy in Early Modern England: The Prothero Lecture, Read 2 July 1986," in *Transactions of the Royal Historical Society*, 5th Series, no. 37 (1987), 118.

28. Schaffer, "Science," 399–400.

29. Van Egmond, *Practical Mathematics*, 6. Anonymous, *An introduction for to lerne to recken with the pen or with the counters* (London: Nycolas Bourman, 1539), A1r.

30. Meskens, "Mathematics Education," 152. Simmons, "ABCs, Almanacs, Ballads," 505.

31. Jean Vanes, *Education and Apprenticeship in Sixteenth-Century Bristol* (Bristol: Bristol Branch of the Historical Association, 1982), 21–2.Travis D. Williams, "The Earliest English Printed Arithmetic Books," *The Library: The Transactions of the Bibliographical Society*, 7th ser., 13, no. 2 (2012): 175.

32. Anonymous, *The arte and science of arismetique* (London: Rychard Fakes, 1526), 1r.

33. For further analysis of the influences, structures, and authorships of these early arithmetic textbooks, see Williams, "Earliest English Printed," 164–84.

34. These new arithmetics included bestsellers by James Hodder, Edmund Wingate, and especially Edward Cocker.

35. Didactic books, in general, often had long "afterlives" in the early modern period. Natasha Glaisyer, "Popular Didactic Literature," in Joad Raymond, ed., *The Oxford History of Popular Print Culture*, vol. 1, *Cheap Print in Britain and Ireland to 1660* (Oxford: Oxford University Press, 2011), 514. .

36. 21793, fol. 2v, Huntington Library, San Marino, CA (HEH).

37. Harkness, *Jewel House*, 118.

38. John Brinsley, *Ludus Literarius: or, the Grammar Schoole; shewing how to proceede from the first entrance into learning, to the highest perfection required in the Grammar Schooles* (London: Felix Kyngston, 1627), 26.

39. Swetz, *Capitalism and Arithmetic*, 24. For more on Recorde's career as a physician and royal administrator, see Jack Williams, *Robert Recorde: Tudor Polymath, Expositor and Practitioner of Computation* (London: Springer, 2011); Jack Williams, "The Lives and Works of Robert Recorde," in Gareth Roberts and Fenny Smith, eds., *Robert Recorde: The Life and Times of a Tudor Mathematician* (Cardiff: University of Wales Press, 2012), 7–24; Taylor, *Mathematical Practitioners*, 15, 167, 313; and *Oxford Dictionary of National Biography*, s.v. "Recorde, Robert (c. 1512–1558), *mathematician*," by Stephen Johnston, http://www.oxforddnb.com/view/article/23241. For more on Baker, see Taylor, *Mathematical Practitioners*, 172, 318; and *Oxford Dictionary of National Biography*, s.v. "Baker, Humphrey (*fl.* 1557–1574), *writer on astrology and arithmetic*," by Anita McConnell, http://www.oxforddnb.com/view/article/1123.

40. For more on the use of dialogue in pedagogical texts, see Peter Burke, "The Renaissance Dialogue," *Renaissance Studies* 3, no. 1 (March 1989): 1–12; Ian Green, *The Christian's ABC: Catechisms and Catechizing in England, c.1530–1740* (Oxford: Clarendon Press, 1996), 17–21; Schaffer, "Science," 401; and Glaisyer, "Popular Didactic Literature," 513. On the blurred line between oral and literate knowledge in early modern England more generally, see Fox, *Oral and Literate Culture*.

41. Robert Recorde, *The Ground of Artes Teachyng the Worke and Practise of Arithmetike* (London: Reynold Wolfe, 1543), 8r–v. The work was officially dedicated to a landowner and royal official named Richard Whalley, who had at least five children of an age to be learning arithmetic at that time. *Oxford Dictionary of National Biography*, s.v. "Whalley, Richard (1498/9–1583), *administrator*," by Alan Bryson, http://www.oxforddnb.com/view/article/29161.

42. Robert Recorde, *The Grounde of Artes*, ed. John Mellis (London: I. Harison and H. Bynneman, 1582), A2v.

43. Robert Recorde, *The Grounde of Artes*, ed. John Mellis and John Wade (London: William Hall, 1610), Mm8r. For more on Mellis, see *Oxford Dictionary of National Biography*, s.v. "Mellis, John (*fl. c.* 1564–1588), *writer on arithmetic and bookkeeping*," by Thompson Cooper, http://www.oxforddnb.com/view/article/18529.,

44. Mellis's accounting textbook was also a newly revised edition of an earlier book, *Profitable Treatyce* by Hugh Oldcastle, but in this case the original is no longer extant. Yamey, Edey, and Thomason, *Accounting*, 155–9. Hugh Oldcastle, *A Briefe Instruction and maner how to keepe bookes of Accompts after the order of Debitor and Creditor, & as well for proper Accompts particle, &c.*, ed. John Mellis (London: Iohn Windet, 1588).

45. Robert Recorde, *The Grounde of Artes*, ed. John Dee, John Mellis, and Robert Hartwell (London: John Beale, 1623), 596. Robert Recorde, *The Ground of Artes*, ed. John Dee, John Mellis, and Robert Hartwell (London: Thomas Harper, 1632), 611.

46. Recorde, *Grounde of Artes* (1623), Rr8v.

47. Harkness, *Jewel House*, 133. For more on early modern instruments, see Jim Bennett, "Early Modern Mathematical Instruments," *Isis* 102, no. 4 (December 2011): 697–705.

48. Cab Lib g, Society of Antiquaries, London.

49. Joseph A. Dane and Alexandra Gillespie, "The Myth of the Cheap Quarto," in John N. King, ed., *Tudor Books and Readers: Materiality and the Construction of Meaning* (Cambridge, 2010), 25–45. Robert Clavell, *A Catalogue of All the Books Printed in England since the Dreadful Fire of London in 1666, to the End of Michaelmas Term, 1672* (London: S. Simmons for R. Clavel, 1673), 43.

50. Simmons, "ABCs, Almanacs, Ballads," 506. Adrian Johns, "Science and the Book," in John Barnard and D. F. McKenzie, eds., *Cambridge History of the Book in Britain*, vol. 4, *1557–1695* (Cambridge: Cambridge University Press, 2002), 289.

51. X513 W72p 1630, University of Illinois, Urbana-Champaign Special Collections, 3r; C.175.d.34, British Library (BL), 2v; 1607/500, BL, 1r; 8532.aa.24, BL, 2r; Adams.8.65.35, Cambridge University Library (CUL), 1r; Vet.A3 f.1247, Bodleian Library, Oxford University, 2r; and 313383, HEH, 1r.

52. Clavell, *Catalogue*, 43.

53. 646/A, Wellcome Library, London; 8506.aa.34, BL; M.6.58, CUL; and C.115.n.43, fol. A2r, BL.

54. Watt, *Cheap Print*, 263. Thomas, "Numeracy," 120.

55. James Hodder, *Hodder's Arithmetick: or, That Necessary Art Made Most Easie* (London: J. Darby, 1667), a4v–a5r.

56. *Oxford Dictionary of National Biography*, s.v. "Hodder, James (*fl.* 1659–1673), arithmetician," by Ruth Wallis, http://www.oxforddnb.com/view/article/13416.

57. John Mayne, *Arithmetick: Vulgar, Decimal, & Algebraical* (London: J. A., 1675), A3r.R. B., *An Idea of Arithmetick* (London: J. Flesher, 1655), A1r.

58. Edward Cocker, *Cocker's Arithmetick* (London: Thomas Passinger, 1678), A2v–A3r.

59. For an example of book ownership versus book reading, see Owen Gingerich, *The Book Nobody Read: Chasing the Revolutions of Nicolaus Copernicus* (New York: Walker, 2004).

60. For an example of a manuscript arithmetic textbook, see MS HA School Exercises Box 5, Folder 1, HEH, which is an educational common-place book dating to 1623. The book also contains extensive notes on geometry and rules of measurement, as presented by the London mathematical tutor, John Speidell. For more on such manuscripts, particularly as they continued to be used into the eighteenth and early nineteenth centuries, see John Dennis, *Figuring It Out: Children's Arithmetical Manuscripts, 1680–1880* (Oxford: Huxley Scientific Press, 2012) and Nerida Ellerton and M. A. (Ken) Clements, *Rewriting the History of School Mathematics in North America, 1607–1861: The Central Role of Cyphering Books* (Dordrecht: Springer, 2012).

61. William Sherman, *Used Books: Marking Readers in the Renaissance England* (Philadelphia: University of Pennsylvania Press, 2008), 3.

62. Sherman, *Used Books*, 5–6, 9. [DeM] L.1 [Cocker] SSR.1700, fol. 3r, SHL.

63. The books in this sample set are not random but are a total population sample of arithmetic textbooks from libraries that I have been able to visit. For an examination of eighteenth-century marginalia, see Benjamin Wardhaugh, "'The Admonition of a Good-Natured Reader': Marks of Use in Georgian Mathematical Textbooks," in Philip Beeley, Yelda Nasifoglu, and Benjamin Wardhaugh, eds., *Reading Mathematics*

in Early Modern Europe: Studies in the Production, Collection, and Use of Mathematical Books (New York: Routledge, 2021): 23051.

64. 512 K47e, fol. K1v, American Philosophical Society Library, Philadelphia.

65. They do have library markings, including notes on acquisition and rebinding, which generally date to the twentieth century. For example, the subset of arithmetic textbooks from the Senate House Library, University of London (SHL), were largely acquired in the nineteenth century by Augustus De Morgan, and most contain notes in his hand.

66. Of the libraries consulted, the three with the largest collections of arithmetic textbooks—the British Library, the Huntington Library, and the Bodleian Library—had overall marginalia rates of 72%, 66%, and 58%, respectively. It is worth noting that the Huntington Library, which now includes the Burndy Library collection and whose holdings Sherman studied extensively, has marginalia rates in line with its two nearest-size neighbors. The next three largest collections—held by the University of Illinois at Urbana-Champaign, the University of London's Senate House Library, and Cambridge University Library—show somewhat higher marginalia rates at 81%, 85%, and 84%, respectively. This higher rate of marginalia also holds for libraries with even smaller collection sizes, where sampling size makes percentages less useful. It is probable that a collecting bias toward "clean" copies of books has affected the numbers for the largest libraries, as opposed to the smallest libraries, where copies were acquired on a more ad hoc basis. Looking at marginalia by decade of publication reveals marginalia in between 50% and 100% of books, but the lower survival rate of textbooks prior to the 1570s is probably responsible for the most extreme percentages. After the 1570s, marginalia rates by publication decade range between 60% and 90%.

67. For more on the breakdown of textbooks by libraries and first author, see the appendix.

68. Edmund Wingate, *Arithmetique Made easie, In Two Bookes* (London: Miles Flesher, 1630), 4v–A1r.

69. Edmund Wingate, *Mr. Wingate's Arithmetick*, ed. John Kersey (London: Philemon Stephens, 1658), A4r–A5r. These changes were likely made possible by Edmund Wingate's death in 1656. Taylor, *Mathematical Practitioners*, 205. *Oxford Dictionary of National Biography*, s.v. "Wingate, Edmund (*bap.* 1596, *d.* 1656), *mathematician and legal writer*," by Bertha Porter, rev. by H. K. Higton, http://www.oxforddnb.com/view/article/29732.

70. BL Add. MSS. 4239, fol. 18r. The inclusion of Hill in this list is curious, as the *ESTC* only records one edition of his arithmetic textbook, as opposed to Recorde and Baker's frequently reprinted textbooks. It is possible that other editions have been lost to the historical record or that Martindale had personal experience with Hill's arithmetic that made him highly value the book despite its failure to be reprinted. Thomas Hylles, *The arte of vulgar arithmeticke* (London: Gabriel Simson, 1600).

71. Adam Martindale, *The Countrey-Survey-Book: or Land-Meters Vade Mecum* (London: A. G. and J. P., 1692), M3r. Martindale was not the only one to begin writing an arithmetic textbook, but he never make it to publication. For example, BL Add. MSS. 4473, fols. 24–7, contains the partially completed textbook of "William Senior professior of the Mathematiques 1641," who taught mathematics out of his house.

72. Humfrey Baker, *The welspring of sciences* (London: Henry Denham, 1564), A1r.

73. Humfrey Baker, *The wel spring of sciences* (London: Thomas Purfoote, 1591), A1r..

74. Robert Recorde, *The Grounde of Artes*, ed. John Dee and John Mellis (London: N. Okes, 1607), A1r.

75. Robert Norton had previously translated a Dutch treatise on decimal arithmetic and published it in 1608. Simon Stevin, *Disme: the Art of Tenths, or, Decimall Arithmetike*, trans. Robert Norton (London: S. Stafford, 1608). Robert Recorde, *The Grounde of Artes*, ed. John Dee, John Mellis, and Robert Norton (London: Thomas Snodham, 1615), A1r.

76. Robert Recorde, *The Grounde of Artes*, ed. John Dee et al. (London: Thomas Harper, 1631), A1r.

77. *OED*, s.v. "vulgar, *adj.*"

78. Humfrey Baker, *Baker's Arithmetick*, ed. Henry Phillippes (London: E. C. & A. C., 1670), A1r.

79. Jonas Moore, *Moores arithmetick: discovering the secrets of that art, in number and species. In two books.* (London: Thomas Harper, 1650), A1r.

80. John Cannon, *Memoirs of the Birth, Education, Life and Death of Mr. John Cannon. Sometime Officer of the Excise & Writing Master at Mere Glastenbury & West Lydford in the County of Somerset*, Somerset Records Office, DD/SAS C/1193/4, 59-60. This source was generously shared with me by Tim Hitchcock.

81. *Oxford Dictionary of National Biography*, s.v. "Cocker, Edward (1631/2–1676), *calligrapher and arithmetician*," by Ruth Wallis, http://www.oxforddnb.com/view/article/5779. See also the auction advertisement pasted inside the back cover of SHL [D.-L.L] L2[Cocker]SR.

82. Edmund Wingate, *Arithmetique Made Easie*, ed. John Kersey (London: J. Flesher, 1650), 4625. *Oxford Dictionary of National Biography*, s.v.v. "Kersey, John, the elder (*bap.* 1616, *d.* 1677), *mathematician*," and "John Kersey the younger (b. c. 1660, d. in or after 1721)," by Ruth Wallis, http://www.oxforddnb.com/view/article/15474.

83. James Hodder, *Hodder's Arithmetick*, ed. Henry Mose (London: Ric. Chiswell and Tho. Sawbridge, 1683), A8v.

84. Martindale, *Countrey-Survey-Book*, M3r.

85. For an excellent, recent discussion of these issues, see Ian Green, *Humanism and Protestantism in Early Modern English Education* (Burlington, VT: Ashgate, 2009).

86. For more on the universities' resistance to any new institutions of learning—regardless of proposed curriculum—that might challenge their supremacy, see Mordechai Feingold, "Tradition versus Novelty: Universities and Scientific Studies in the Early Modern Period," in *Revolution and Continuity: Essays in the History and Philosophy of Early Modern Science*, ed. Peter Barker and Roger Ariew (Washington, DC: Catholic University of America Press, 1991), 45–59.

87. Feingold, *Mathematicians' Apprenticeship*. See also Mordechai Feingold, "Reading Mathematics in the English Collegiate-Humanist Universities." In *Reading Mathematics in Early Modern Europe: Studies in the Production, Collection, and Use of Mathematical Books*, edited by Philip Beeley, Yelda Nasifoglu, and Benjamin Wardhaugh, 124–50. New York: Routledge, 2021..

88. Charles Hoole, *The Petty-Schoole, Shewing a Way to Teach Little Children to Read English with Delight and Profit, (especially) According to the New Primar* (London: J. T. for Andrew Crook, 1659), 2.

89. Helen M. Jewell, *Education in Early Modern England* (New York: St. Martin's Press, 1998), 17.

90. The success of at least some dame schools and the informal teaching of arithmetic by other women can be inferred by the evidence of women managing their own and their family's accounts in the early modern period. See, for example, Susan E. James, *Women's Voices in Tudor Willis, 1485–1603* (New York: Routledge, 2016), 100, 115, 118, 201, 209–10.

91. Richard DeMolen, *Richard Mulcaster and Educational Reform in the Renaissance* (Nieuwkoop: De Graaf, 1991), xviii.

92. *ESTC*, record number 006176804.

93. Francis Clement, *The Petie Schole with an English Orthographie* (London: Thomas Vautrollier, 1587), A1r.

94. Clement, *The Petie Schole*, 65, 71–85.

95. Hoole, *Petty-Schoole*, 30.

96. For more on Rechenmeister counters, see Chapter 2.

97. Schoolmasters' licenses rarely survived in full, as they were kept by individual schoolmasters in their private records. Instead, most instances of licenses come from ecclesiastical visitations, where the contents of licenses were summarized for the visitation record. David Cressy, *Education in Tudor and Stuart England* (New York: St. Martin's Press, 1975), 32.

98. In this usage, "accidence" signifies the "branch of grammar which deals with the inflection of words, grammatical morphology," OED, s.v. "accidence, n2." Cressy, *Education*, 33–4.

99. Cressy, *Literacy*, 36.

100. Cressy, *Literacy*, 35–41.

101. Christopher Wase, *Considerations Concerning Free-Schools as Settled in England* (London: Simon Millers, 1678), 33.

102. Kenneth Charlton, *Women, Religion and Education in Early Modern England* (London: Routledge, 1999), 146.

103. Jewell, *Education*, 95.

104. John Lawson and Harold Silver, *A Social History of Education in England* (London: Routledge, 1973), 107. Robert Ashton, *Oxford Dictionary of National Biography*, s.v. "Leman, Sir John (1544–1632), *merchant and mayor of London*," by Robert Ashton http://www.oxforddnb.com/view/article/16420. Charlton, *Women, Religion and Education*, 151, 148. N. Plumley, "The Royal Mathematical School within Christ's Hospital: The Early Years—Its Aims and Achievements," *Vistas in Astronomy* 20 (1976): 58.

105. Charlton, *Women, Religion and Education*, 147.

106. Cressy, *Literacy*, 30. Charlton, *Women, Religion and Education*, 149.

107. Edmund Coote, *The English School-Master* (London: B. Alsop, 1651), A2r, H2r.

108. See, for example, Anonymous, *The ABC with The Catechisme: That is to say, an Instruction to bee taught and learned of euery Child, before he be brought to be*

confirmed by the Bishop (London: n.p., 1633), which was reprinted at various times throughout the seventeenth century.

109. John White, *The Country-Man's Conductor in reading and writing true English . . . and some arithmetical rules to be learnt by children, before or as soon as they are put to Writing* (Exeter: Samuel Farley, 1701), A1r, A5v.

110. White, *Country-Man's Conductor*, A1r.

111. Thomas, "Meaning of Literacy," 102–3.

112. Jewell, *Education*, 856.

113. Brinsley, *Ludus Literarius*, 25.

114. Jewell, *Education*, 84.

115. Although the grammar school was teaching Arabic numerals and ciphering as early as 1597, the school's various accountants used Roman numerals to record monetary entries and sums until 1669/70. George Alfred Stocks, ed., *The Records of Blackburn Grammar School*, Remains, Historical and Literary, connected with the Palatine Counties of Lancashire and Chester, n.s., 66 (Manchester: Charles E. Simms, 1909), 1:73.

116. Stocks, *The Records of Blackburn Grammar* School, 1:74.

117. Cressy, *Education*, 65.

118. William Lilly, *Merlini Anglici Ephemeris: Or, Astrological Judgments for the year 1677* (London: J. Macock, 1677), F8v.

119. Mellis advertised his school in the versions of Recorde's *The Ground of Artes* that he edited, from 1582 until 1607. His advertisement also appeared in the 1610 edition, "now lastly corrected by John Wade," but was replaced by N. Physhe in the 1615 edition. Recorde, *Ground of Artes* (1607), Mm8r; Recorde, *Ground of Artes* (1610), A1r; Recorde, *Records Arithmeticke* (1615), Oo3v.

120. *Oxford Dictionary of National Biography*, s.v. "Hodder, James (*fl.* 1659–1673), arithmetician," by Ruth Wallis, http://www.oxforddnb.com/view/article/13416.

121. *Oxford Dictionary of National Biography*, s.v. "Cocker, Edward (1631/2–1676), calligrapher and arithmetician," by Ruth Wallis, http://www.oxforddnb.com/view/article/5779.

122. Edward Cocker, *Cocker's Arithmetick*, ed. John Hawkins (London, 1680).

123. Lilly, *Merlini Anglici Ephemeris*, F8v.

124. Sarah Powell and Paul Dingman, "Arithmetic Is the Art of Computation," *The Collation* (blog), September 8, 2015, http://collation.folger.edu/2015/09/arithmetic-is-the-art-of-computation.

125. Green, *Humanism*, 310.

126. Chris Minns and Patrick Wallis, "Rules and Reality: Quantifying the Practice of Apprenticeship in Early Modern England," *Economic History Review* 65, no. 2 (May 2012): 559. Patrick Wallis, "Apprenticeship and Training in Premodern England," *Journal of Economic History* 68, no. 3 (September 2008): 836.

127. Arthur J. Willis and A. L. Merson, eds., *A Calendar of Southampton Apprenticeship Registers, 1609–1740* (Southampton: Southampton University Press, 1968), 19.

128. John Rigges, apprenticed in 1611 to his uncle, was to be instructed in his uncle's trade and "alsoe to be enxtructed in all other trades or sciences as the said Frauncis Rigges shall use during the said terme." Willis and Merson, *Calendar*, 2.

129. Vanes, *Education and Apprenticeship*, 21.
130. Willis and Merson, *Calendar* 86.
131. Willis and Merson, *Calendar*, 38.

Chapter 4

1. Arthur J. Willis and A. L. Merson, eds., *A Calendar of Southampton Apprenticeship Registers, 1609–1740* (Southampton: Southampton University Press, 1968), 1.
2. In the scholarly literature, this multiplicity has been referred to as pluritemporalism, multiple temporal coexistence, and/or a mixture of times. Keith Wrightson, "Popular Senses of Past Time: Dating Events in the North Country, 1615–1631," in Michael J. Braddick and Phil Withington, eds., *Popular Culture and Political Agency in Early Modern England and Ireland: Essays in Honor of John Walter* (Woodbridge, Suffolk: Boydell and Brewer, Boydell Press, 2017), 100. Paul Glennie and Nigel Thrift, *Shaping the Day: A History of Timekeeping in England and Wales 1300–1800* (Oxford: Oxford University Press, 2009), 66.
3. Glennie and Thrift, *Shaping the Day*, 226. The date of Easter was notoriously difficult to calculate, and a formula was not derived for it until 1876. Before then, a series of complex calculations were used to determine the date. Robert W. Poole, *Time's Alteration: Calendar Reform in Early Modern England*, (London: UCL Press, 1998), 33–5.
4. Since 1969, Catholics now celebrate this feast on July 3.
5. See especially definition A.I.1.a. for "time" in the *OED*, which states that time is a "finite extent or stretch of continued existence, as the interval separating two successive events or actions, or the period during which an action, condition, or state continues." *OED*, s.v. "time, *n., int.*, and *conj.*"
6. Glennie and Thrift, *Shaping the Day*, 223.
7. For more on how calendars evolved to help predict future events, see Johan De Smedt and Helen De Cruz, "The Role of Material Culture in Human Time Representation: Calendrical Systems as Extensions of Mental Time Travel," *Adaptive Behavior* 19, no. 1 (2011): 63–76, https://doi.org/10.1177/1059712310396382.
8. Alexander Philip, *The Calendar: Its History, Structure and Improvement* (Cambridge: Cambridge University Press, 1921), 83.
9. This calendar system was an amalgamation of various cultures' dating practices as adopted and disseminated by the Roman Empire. The month names came from Greece but were modified by various Roman emperors. The traditional Roman week had eight days but was replaced with the Jewish and Christian seven-day week in the first century AD. The twenty-four-hour day came from Egypt, while the less-used cycle of sixty minutes and seconds originated in Mesopotamia. Eviatar Zerubavel, *The Seven Day Circle: The History and Meaning of the Week* (Chicago: University of Chicago Press, 1989), 16, 20, 23. Ken Mondschein, *On Time: A History of Western Timekeeping* (Baltimore: Johns Hopkins University Press, 2020), 24. Jennifer Powell

McNutt, "Hesitant Steps: Acceptance of the Gregorian Calendar in Eighteenth-Century Geneva," *Church History* 75, no. 3 (2006), 545.

10. Clock hours were regularly divided into halves, thirds, or quarters in everyday life, but not into individual minutes or seconds. Gerhard Dohrn-van Rossum, *History of the Hour: Clocks and Modern Temporal Orders*, trans. Thomas Dunlap (Chicago: University of Chicago Press, 1996), 282.

11. The Julian reforms themselves were in response to the calendar being so out of alignment with astronomical observations that the year 46 BC was extended to 445 days in length and dubbed the "year of confusion." Poole, *Time's Alteration*, 32.

12. This is the tropical version of the solar year. Another possible way to calculate the solar year is according to the length of time it takes for the sun to return to the same place among a set of constellations—the sidereal version of the solar year. Neither of these methods for calculating the solar year presupposes a heliocentric model of the solar system. Other calendars are based on the lunar year, while the Hebrew calendar is lunisolar—accommodating both lunar and solar years.

13. There was a great deal of confusion when the leap year was first instituted, and for several years the leap day was added to every third year instead of every fourth. Consequently, the calendar had to be adjusted again by Augustus Caesar, who caused the leap day to be dropped from 5 BC, 1 BC, and 4 AD. Augustus also reformed the Egyptian, Alexandrian calendar to synchronize with the Roman, Julian calendar. Walter F. Snyder, "When Was the Alexandrian Calendar Established?" *American Journal of Philology* 64, no. 4 (1943): 387. T. C. Skeat, "The Egyptian Calendar under Augustus," *Zeitschrift für Papyrologie und Epigraphik*, Bd. 135 (2001): 153–4.

14. Despite the cycle of hours officially restarting half an hour after sunset, events that occurred overnight were commonly considered to have occurred on the previous "day." Rossum, *History of the Hour*, 114. Glennie and Thrift, *Shaping the Day,* 27.

15. Hence the reason that the months of September, October, November, and December bear the prefixes for the numbers 7 through 10, but are no longer the seventh, eighth, ninth, and tenth months of the year, respectively.

16. Kristen Poole and Owen Williams, eds., *Early Modern Histories of Time: The Periodizations of Sixteenth- and Seventeenth-Century England* (Philadelphia: University of Pennsylvania Press, 2019), 6.

17. The method of reckoning the year from the incarnation of Christ was standardized in the eighth century by the Venerable Bede, who used chronicles and the gospels to establish the dates of creation and the incarnation. Poole, *Time's Alteration*, 35.

18. The BC/AD distinction was made by Bede, in his *Ecclesiastical History of the English Nation*, but it was not used by other scholars until the fifteenth century and did not come into widespread use until the seventeenth century. G. J. Whitrow, *Time in History: The Evolution of Our General Awareness of Time and Temporal Perspective* (Oxford: Oxford University Press, 1988), 70. The inherently religious foundation of still widely used BC/AD time reckoning has led some scholars to prefer the terminology BCE/CE (Before Common Era/Common Era) but the temporal origin point remains the same.

19. Arno Borst, *The Ordering of Time: From the Ancient Computus to the Modern Computer*, trans. Andrew Winnard (Chicago: University of Chicago Press, 1993), 87.

20. This was an extremely complicated calculation. While the lunar and solar cycles coincide every nineteen years, that does not take into account the requirement that Easter take place on a Sunday. Because of the leap year, it takes twenty-eight years for the solar cycle and the weekly cycle to coincide. Thus all three cycles—weekly, lunar, and solar—coincide every 532 years. Donald J. Wilcox, *The Measure of Times Past: Pre-Newtonian Chronologies and the Rhetoric of Relative Time* (Chicago: Chicago University Press, 1987), 133.

21. As Edward Chamberlayne complained, "yet cannot it be denied but that this old Computation is become erroneous; for by our Rules, two *Easters* will be observed within one year, as in the last 1667, and not one *Easter* to be observed this year," that is, 1668. Edward Chamberlayne, *Angliae Notitia, or The Present State of England* (London: T. N. for John Martyn, 1669), 70. Or, as an anonymous wit remarked more pithily, "He'd rather have two *Easters* in a Year, / Than to disturb the *sacred Calendar*." Anonymous, *An Account of the solemn reception of Sir Iohn Robinson, Lord-Maior at St. Pauls Cathedral, the day of his inauguration* (London: Josuah Coniers,1662), 7.

22. McNutt, "Hesitant Steps," 546.

23. Medieval astronomers and philosophers who advocated calendar reform included Roger of Hereford in the twelfth century, and Robert Grosseteste, Jean Sacrobosco, and Roger Bacon in the thirteenth century. Poole, *Time's Alteration*, 36.

24. Aside from the wholesale deletion of ten days, the Gregorian calendar retained almost all the same features as the Julian calendar. This included all the Julian chronological units, with only a slight modification to the Julian month of February: February 29 was to be skipped in years divisible by 100 that were not also divisible by 400.

25. The Eastern Orthodox churches instituted their own reform of the Julian calendar in the twentieth century.

26. Although Catholics were more likely to adopt the reform, while Protestants rejected it, this was not universal. Catholic France adopted the calendar late and skipped from December 9 to 20. The Protestant Low Countries, under the control of the French Duke of Alençon, also adopted the new calendar in December. Other Catholic polities adopted it at various points in the mid-to-late 1580s, well after the date set by the papal bull. Poole, *Time's Alteration*, 39, 46. Anne Lake Prescott, "Refusing Translation: The Gregorian Calendar and Early Modern English Writers," *Yearbook of English Studies* 36, no. 1 (2006): 8. For more on Continental Protestants' reactions to calendar reform, particularly the German *kalenderstreit*, see Poole, *Time's Alteration*, 38–9; Rona Johnston Gordon, "Controlling Time in the Hapsburg Lands: The Introduction of the Gregorian Calendar in Austria Below the Enns," *Austrian History* Yearbook 40 (2009): 28–36;James Hodgson, *An Introduction to Chronology* (London: J. Hinton, 1747); McNutt, "Hesitant Steps," 544–64.

27. For more on the failed 1583 calendar reform, and the roles Elizabeth, her Secretary of State Francis Walsingham, John Dee, and other state and religious officials played

in the reform attempt, see Glyn Parry, *The Arch Conjuror of England: John Dee* (New Haven, CT: Yale University Press, 2012), 146–61.

28. John Dee, *A Playne Discourse . . . concerning the nedful Reformation of the Vulgar Kalendar for the civile years and daies accompting, or verifyeng, according to the time truely spent*, as cited in Poole, *Time's Alteration*, 47.

29. Poole, *Time's Alteration*, 59–60.

30. Glennie and Thrift, *Shaping the Day*, 277n8. Poole, *Time's Alteration*, 48. Alison A. Chapman, "The Politics of Time in Edmund Spenser's English Calendar," *Studies in English Literature, 1500–1900* 42, no. 1 (2002): 6. Prescott, "Refusing Translation," 2. SP 12/160 f.62 (Gale Document Number MC4304281096).

31. Poole, *Time's Alteration*, 49–51. Parry, *Arch Conjuror*, 158–9.

32. Poole, *Time's Alteration*, 49–51. Parry, *Arch Conjuror*, 158–9.

33. Poole, *Time's Alteration*, 51.

34. Poole, *Time's Alteration*, 51.

35. A modern analogy might be someone who can use the power rule and chain rule to find a derivative but doesn't know or understand the fundamental theorem of calculus.

36. SP 83/17 f.174 (Gale Document Number MC4312400532).

37. SP 94/1 f.109 (Gale Document Number MC4312400526); SP 83/17 f.185 (Gale Document Number MC4312400538); SP 83/18 f.29 MC 4312500051; SP 81/2 f.129 (Gale Document Number MC4312500124).

38. SP 83/18 f.46 (Gale Document Number MC4312500085); SP 101/27 f. 3 (Gale Document Number MC4312500713); SP 83/18 f.39 (Gale Document Number MC4312500070); SP 83/17 f.183 (Gale Document Number MC4312400537).

39. SP 84/31 f.180 (Gale Document Number MC4313400286). Nadine Akkerman, ed. *The Correspondence of Elizabeth Stuart, Queen of Bohemia*, vol. 2, *1632–1642* (Oxford: Oxford University Press, 2011), 74, 467.

40. Humfrey Baker, *The vvell spryng of sciences* (London: Ihon Kyngston, 1568), G3r.

41. SP 92/2 f. 319 (Gale Document Number MC4891180134).

42. Akkerman, *Correspondence of Elizabeth Stuart*, 20, 69, 183.

43. Akkerman, *Correspondence of Elizabeth Stuart*, 777.

44. Akkerman, *Correspondence of Elizabeth Stuart*, 118, 813, 899.

45. For examples, see Akkerman, *Correspondence of Elizabeth Stuart*, 27, 60, 74, 163–4, 467, 694, 997.

46. Akkerman, *Correspondence of Elizabeth Stuart*, 1004.

47. See, for example, the dates in James Croston, ed., *The Register Book of Christenings, Weddings, and Burials within the Parish of Prestbury, in the County of Chester, 1560–1636*, Record Society of Lancashire and Cheshire 5 (Manchester: A. Ireland and Co., 1881); Carol G. Bonsey and J. G. Jenkins, eds., *Ship Money Papers and Richard Grenville's Note-Book*, Buckinghamshire Record Society Publications 13, (Hertfordshire: Broadwater Press, 1965); and J. P. Earwaker, ed., *Lancashire and Cheshire Wills and Inventories at Chester*, Remains, Historical and Literary, connected with the Palatine Counties of Lancaster and Chester, n.s., 3 (Manchester: Charles E. Simms, 1884), 242.

48. Poole, *Time's Alteration*, 112.
49. R. W. Greaves, ed., *The First Ledger Book of High Wycombe*, Buckinghamshire Record Society Publications 11 (Hertfordshire: Broadwater Press, 1947), 108.
50. BL, MS Additional 28004, ff. 197r. Willis and Merson, *Southampton*, 48.
51. Akkerman, *Correspondence of Elizabeth Stuart*, 34, 399.
52. While both January 1 and March 25 were red letter days—printed in red, rather than black, ink to call attention to the holiday—the former was generally printed as "New Years Day" while the latter was referred to as the "Annunciation" or the "Annunciation of Mary."
53. *OED*, s.v. "almanac, *n.*"
54. Bernard Capp, *Astrology and the Popular Press: English Almanacs 1500–1800* (London: Faber and Faber, 2008), 29. This is a reprint of the original 1979 edition from Cornell University Press.
55. Capp, *Astrology and the Popular Press*, 41, 44.
56. Capp, *Astrology and the Popular Press*, 44. In other ways, almanacs came first. Johannes Gutenberg printed an almanac on his new press, even before his more famous Bible. John Durham Peters, "Calendar, Clock, Tower," in Jeremy Stolow, ed., *Deus in Machina: Religion, Technology, and Things in Between* (New York: Fordham University Press, 2013), 28.
57. For examples of almanacs discussing calendar reform, see, for example, Arthur Hopton, *Hopton 1612. An Almanack and Prognostication* (London: Company of Stationers, 1612), B3r–B4v; and Francis Perkins, *A New Almanack and Prognostication for the Year of Our Lord God 1671* (London: E. L. and Robert White, 1671), A1r. For an extensive treatment of medical information in almanacs, see Louise Hill Curth, *English Almanacs, Astrology and Popular Medicine: 1550–1700* (Manchester: Manchester University Press, 2007).
58. Capp, *Astrology and the Popular Press*, 52. See, for example, Thomas Bretnor's claim to teach arithmetic, geometry, navigation, astronomy and more in English, Latin, French, or Spanish. Thomas Bretnor, *Bretnor. 1616.* (London: Company of Stationers, 1616), A3v.
59. Glennie and Thrift, *Shaping the Day*, 247.
60. Arthur Hopton, *Hopton 1611 An Almanack and Prognostication* (London: Company of Stationers, 1611), A1v.
61. Alison A. Chapman, "Marking Time: Astrology, Almanacs, and English Protestantism," *Renaissance Quarterly* 60 (2007): 1269–70.
62. Prescott, "Refusing Translation," 3–4.
63. Poole, *Time's Alteration*, 105.
64. Poole and Williams, *Early Modern Histories of Time*, 7.
65. Wrightson, "Popular Senses of Past Time," 95–7.
66. Wrightson, "Popular Senses of Past Time," 101–2.
67. Easter term took its name from its Easter-determined starting date, while Trinity term began the day after Corpus Christi day. As Richard Allestree explained to his readers, "Note that Corpus Christi day is alwayes the Thursday after Trinity Sunday." Richard Allestree, *Allestree. 1641. A New Almanack and Prognostication* (London: T. Coates, 1641), A2r.

68. George Wharton, *No Merline, nor Mercury: but a New Almanack after the Old Fashion* (London, 1648), A5r.

69. This same mentality can be seen in modern references to holiday names rather than their Gregorian months and days; US stores advertise their Veteran's Day sales rather than their November 11 sales. For more on early modern holidays and festival culture, see David Cressy, *Bonfires and Bells: National Memory and the Protestant Calendar in Elizabethan and Stuart England* (Berkeley: University of California Press, 1989); and Ronald Hutton, *The Stations of the Sun: A History of the Ritual Year in Britain* (Oxford: Oxford University Press, 1996).

70. Poole, *Time's Alteration*, 122–4.

71. Capp, *Astrology and the Popular Press*, 62.

72. Although Sir Edward Don could have conceivably had someone to keep his books for him, internal evidence such as the use of first-person pronouns indicate that the book was written by Don himself. Ralph A. Griffiths, ed., *The Household Book (1510–1551) of Sir Edward Don: An Anglo-Welsh Knight and His Circle*, Buckinghamshire Record Society Publications 33 (Amersham, Buckinghamshire: Buckinghamshire Record Society, 2004), 1, 126.

73. Joan W. Kirby, ed., *The Plumpton Letters and Papers*, Camden Fifth Series 8 (Cambridge: Cambridge University Press, 1996), 220–2.

74. This does not appear to be an artifact of the particular churchwardens recording each account, as there is crossover between churchwardens and different methods of reckoning the year. It is possible that this was done in protest, to avoid dating documents by Mary's regnal years; however, the practice continued for several years after Elizabeth's accession. F. W. Weaver and G. N. Clark, eds., *Churchwardens' Accounts of Marston, Spelsbury, Pyrton*, Oxfordshire Records Series 6 (Oxford: Oxfordshire Record Society, 1925), 68–77.

75. Wrightson finds only three instances of use of regnal years in his sample. Wrightson, "Popular Senses of Past Time," 96.

76. C. H. Firth, *Acts and Ordinances of the Interregnum, 1642–1660* (London: H. M. Stationery Office, 1911), 1262–3. The importance of regnal years in establishing royal authority can also be seen on the Isle of Man, which used its own regnal years in the fifteenth century and pointedly used *anno domini* years without English regnal years in the sixteenth century. Tim Thornton, "Lordship and Sovereignty in the Territories of the English Crown: Sub-kingship and Its Implications, 1300–1600," *Journal of British Studies* 60, no. 4 (October 2021), 852.

77. Poole, *Time's Alteration*, 75. A similar system still calculates American regnal years from "the Independence of the United States of America." See, e.g., Joseph R. Biden Jr., "A Proclamation on the Public Service Recognition Week, 2022," April 29, 2022, https://www.whitehouse.gov/briefing-room/presidential-actions/2022/04/29/a-proclamation-on-public-service-recognition-week-2022/.

78. Samuel Rawson Gardiner, *The Constitutional Documents of the Puritan Revolution 1625–1600* (Oxford: Clarendon Press, 1906), 467.

79. This is approximately a 1% error rate.

80. Willis and Merson, *Southampton*, lxxxi, 24–26, 38, 46, 49, 55, 78, 101, 107.

81. The regnal years of Mary and Philip had a similar construction, as Mary married Philip after she ascended the throne of England. Their regnal years thus also had different starting dates.

82. As Charles I ascended to the thrones of England and Scotland simultaneously, his regnal years were the same in both kingdoms.

83. Willis and Merson, *Southampton*, 19.

84. Capp, *Astrology and the Popular Press*, 30.

85. The Zodiac Man was an image that explained the relationship of the heavens to parts of the human body.

86. Edward Pond, *Pond's Almanack for the Yeare of Our Lord God 1641* (Cambridge: Roger Daniel, 1641), A2r. Also quoted in Hill Curth, *English Almanacs*, 120.

87. See, for example, Daniel Browne, *Browne. 1625. A New Almanack and Prognostication* (London: Company of Stationers, 1625), B8v.

88. See, for example, John Gadbury, *Ephemeris: or, A Diary Astronomical, Astrological, Meteorological, for the Year of Grace 1671* (London: J. C., 1671), A8v.

89. Perkins, *Almanack and Prognostication*, A1r.

90. Gadbury, *Ephemeris: 1671*, A8v. Perkins, *Almanack and Prognostication*, A1r. Thomas Bretnor, *Bretnor 1609 a new almanacke and prognostication* (London: Company of Stationers, 1609), A1v. George Wharton, *Calendarium ecclesiasticum: or A New Almanack after the Old Fashion* (London: John Grismond, 1657), A2r. Henry Coley, *Nuncius Coelestis: or Urania's Messenger* (London: J. Grover, 1679), A1r.

91. Coley, *Nuncius Coelestis*, A1r.

92. See, for example, HEH, 30925 and HEH, 30067, A4r–A8r.

93. Capp, *Astrology and the Popular Press*, 30.

94. Chapman, "Marking Time," 1282.

95. Capp, *Astrology and the Popular Press*, 30.

96. Glennie and Thrift, *Shaping the Day*, 26.

97. David S. Landes, *Revolution in Time: Clocks and the Making of the Modern World*, 2nd ed. (Cambridge, MA: Belknap Press of Harvard University Press, 2000), 48. Rossum, *History of the Hour*, 126, 135. While there were watches in England as early as 1580, they were rare and expensive status pieces. Carlo M. Cipolla, *Clocks and Culture, 1300–1700* (New York: W. W. Norton, 1978), 66.

98. Glennie and Thrift, *Shaping the Day*, 154, 176. Given the near impossibility of synchronizing the bells of early modern clocks, it is unsurprising that multi-parish towns were disincentivized from maintaining multiple parish clocks. Landes, *Revolution in Time*, 79.

99. Glennie and Thrift, *Shaping the Day*, 143, 146–7.

100. Borst, *The Ordering of Time*, 94–5.

101. Glennie and Thrift, *Shaping the Day*, 124, 128.

102. Mark Hailwood, "Time and Work in Rural England, 1500–1700," *Past & Present* 248, no. 1 (2020): 118.

103. Wrightson, "Popular Senses of Past Time," 105.

104. Hailwood, "Time and Work," 94–6.

105. Hailwood, "Time and Work," 98.

106. Rossum, *History of the Hour*, 114. Glennie and Thrift, *Shaping the Day*, 27.

107. Rossum, *History of the Hour*, 117. While telling time was mostly an auditory experience, some clocks did have dials that could be read visually. These dials focused on hours and, in England, usually tracked 12 hours before repeating, though 24-, 8-, and 6-hour dials were also made. Regardless, the dial faces privileged Roman numerals; whenever Arabic numerals coexisted with the Roman on a dial, the Arabic numerals were inevitably smaller. Jane Desborough, *The Changing Face of Early Modern Time, 1550–1770* (Cham, Switzerland: Palgrave Macmillan, 2019), 50–1, 73–4.

108. Glennie and Thrift, *Shaping the Day*, 247, 264.

109. Henry Alleyn, *Alleyn 1607. A double Almanacke & Prognostication* (London: Company of Stationers, 1607), A1r.

110. Henry Seaman, *Kalendarium Nauticum: The Sea-man's Almanack* (London: T. N., 1677), A1r. Hill Curth, *English Almanacs*, 47–8.

111. Chapman, "Marking Time," 1263.

112. Arthur Hopton, *Hopton. 1606. An almanack and prognostication for this the second yeare after leape yeare* (London: Company of Stationers, 1606), A2r.

113. *Oxford Dictionary of National Biography*, s.v. "Moxon, Joseph (1627–1691), *printer and globe maker*," by D.J. Bryden, http://oxforddnb.com/view/article/19466. Joseph Moxon, *A Tutor to Astronomie and Geographie* (London, 1659), 62–3.

114. John Seller, *Jamaica Almanack* (London: 1684). John Gadbury, *Ephemeris: or, a Diary Astronomical & Astrological* (London: J. D., 1677), A1r, C8r.

115. Gadbury, *Ephemeris*, 1677: C8r.

116. Richard Baxter, *The Right Method for a Settled Peace of Conscience, and Spiritual Comfort in 32 Directions* (London: T. Underhil, F. Tyton, and W. Raybould, 1653), 373.

117. William Prynne, *A Briefe Polemicall Dissertation, Concerning the True Time of the Ichoation and Determination of the Lordsday-Sabbath* (London: T. Mabb, 1654).

118. Thomas Wilkinson, *Wilkinson, 1657. A Kelander and Prognostication*, (London: G. Dawson, 1657), C8v.

119. *OED*, s.v.v. "fortnight, *n.*" "sennight, *n.*"

120. Thomas Shepard, *Theses Sabbaticae, or, The Doctrine of the Sabbath* (London, 1650), 50.

121. Poole, *Time's Alteration*, 24.

122. John Goad, *Astro-meteorologica, or, Aphorisms and Discourses of the bodies coelestial* (London, 1686), 94.

123. Wilkinson, *Wilkinson, 1657*, C8v.

124. Genesis 1:5 (KJV).

125. Prynne, *Polemicall Dissertation*, 20.

126. Shepard, *Theses Sabbaticae*, 182.

127. Shepard, *Theses Sabbaticae*, 182.

128. Unlike the Gregorian reforms, Shepard's calendar adjustment alters the cycle of the days of the week, a point he does not address in his comparisons.

129. Shepard, *Theses Sabbaticae*, 182.

130. Poole, *Time's Alteration*, 31.

131. Poole, *Time's Alteration*, 112.

132. Poole, *Time's Alteration*, 113–4. For an extensive debunking of the English calendar riots myth and discussion of the myth's origin, see Poole's *Time's Alteration*, particularly chap. 1.

133. After the 1707 union of England and Scotland, the newly constituted Great Britain continued to use England's March 25 new year's day in official state papers.

Chapter 5

1. L. G. Wickham Legg, ed., *A Relation of A Short Survey of 26 Counties Observed in a seven weeks Journey begun on August 11, 1634 By a Captain, a Lieutenant, and an Ancient All three of the Military Company in Norwich* (London: F. E. Robinson, 1904), 1, 41–2. My thanks to Anne Throckmorton for bringing this source to my attention.

2. The *OED* defines "gaming" as "the action or habit of playing at games of chance for stakes; gambling." The term "gaming" was prevalent throughout the sixteenth and seventeenth centuries, while the currently used term "gambling" did not appear in print until 1726. *OED*, s.vv. "gambling, *n*." "gaming, *n*."

3. For more early modern mathematics and probability, see Ian Hacking, *The Emergence of Probability: A Philosophical Study of Early Ideas About Probability Induction and Statistical Inference*, 2nd ed. (Cambridge: Cambridge University Press, 2006); and Barbara Shapiro, *Probability and Certainty in Seventeenth-Century England: A Study of the Relationships Between Natural Science, Religion, History, Law, and Literature* (Princeton, NJ: Princeton University Press, 1983). For late seventeenth- and eighteenth-century mathematical probability, see Lorraine Daston's *Classical Probability in the Enlightenment* (Princeton, NJ: Princeton University Press, 1998).

4. For more on early modern gaming, see Caroline Baird, *Games and Gaming in Early Modern Drama: Stakes and Hazards* (Cham, Switzerland: Palgrave Macmillan, 2020); and Nicholas Barry Tosney, "Gaming in England, c. 1540–1760" (PhD diss., University of York, 2008).

5. John Ashton, *The History of Gambling in England* (Chicago: Herbert S. Stone, 1899), 12. Baird, *Games and Gaming*, 23. The British Museum has two English dice in its online collection dated to the "Iron Age" immediately before the Roman conquest. "Dice," britishmuseum.org/collection/object/H_1976-0501-50 and britishmuseum. org/collection/object/H_1976-0501-51.

6. Baird, *Games and Gaming*, 3–4, 21.

7. Catherine Perry Hargrave, *A History of Playing Cards and a Bibliography of Cards and Gaming* (Boston: Houghton Mifflin, 1980), 169. Gina Bloom, *Gaming the Stage: Playable Media and the Rise of English Commercial Theatre* (Ann Arbor: University of Michigan Press, 2018), 25–6. Baird, *Games and Gaming*, 3, 38.

8. Baird, *Games and Gaming*, 39. Tosney, "Gaming in England," 24. Bloom, *Gaming the Stage*, 25.

9. 3 Ed. 4, c. 4. Joseph Keble, *The Statutes at Large in Paragraphs and Sections or Numbers, from Magna Charta Until This Time* (London: John Bill, Thomas Newcomb, Henry Hills, Richard Atkins, and Edward Atkins, 1681), 278–9.

10. At between 1.25*p*. and 3*p*. a deck in the middle of the seventeenth century, they were quite affordable. Tosney, "Gaming in England," 27, 29, 32, 194.

11. For an example of the diversity of stakes in children's bets, see Jean Vanes, *Education and Apprenticeship in Sixteenth-Century Bristol* (Bristol: Bristol Branch of the Historical Association, 1982), 10.

12. Tosney, "Gaming in England," 79–80. Ed. S. Taylor, *The History of Playing Cards with Anecdotes of their use in Conjuring, Fortune-Telling and Card-Sharping* (London: [Jo] hn Camden Hotten, Piccadilly, 1865), 103. Keble, *Statutes at Large*, 542–3.

13. Baird, *Games and Gaming*, 40–41. Tosney, "Gaming in England," 129–132.

14. For more on the lottery, see David Dean, "Elizabeth's Lottery: Political Culture and State Formation in Early Modern England," *Journal of British Studies* 50, no. 3 (2011): 587–611, doi:10.1086/659829. On lotteries more generally, see also Daston, *Classical Probability*, 141–9. "The Worshipful Company of Makers of Playing Cards," https://makersofplayingcards.org. There are some digitized images of early modern cards on their website at http://www.playingcardmakerscollection.co.uk/PIC-E.html including one deck that narrates the defeat of the Spanish Armada, http://www.play ingcardmakerscollection.co.uk/Cardhtml/W0432.html.

15. Thomas Gataker, *Of the Natvre and Vse of Lots; A Treatise Historicall and Theologicall* (London: Edward Griffin, 1619), 205–6.

16. Tosney, "Gaming in England," 83–4.

17. Hacking, *The Emergence of Probability*, 4.

18. Daston, *Classical Probability*, 8. Keith Devlin, *The Unfinished Game: Pascal, Fermat, and the Seventeenth-Century Letter that Made the World Modern* (New York: Basic Books, 2008), 8. F. N. David, *Games, Gods and Gambling: The Origins and History of Probability and Statistical Ideas from the Earliest Times to the Newtonian Era* (New York: Hafner, 1962), 55. Oystein Ore, *Cardano the Gambling Scholar* (Princeton, NJ: Princeton University Press, 1953), 205.

19. Ashton, *History of Gambling*, 16. Sir Nicholas L'Estrange lived from 1604 to 1655. Chris R. Kyle, "L'Estrange, Sir Nicholas, first baronet (*bap.* 1604, *d.* 1655), collector of anecdotes," in *ODNB* (Oxford: Oxford University Press, 2005), https://doi.org/10.1093/ref:odnb/16513.

20. *OED*, s.v. "odds, *n.*"

21. For a longer discussion of the evolutions of the term "odds" during the early modern period, see Jessica Marie Otis, "'Sportes and Pastimes, done by Number': Mathematical Games in Early Modern England," in Allison Levy, ed., *Playthings in Early Modernity: Party Games, Word Games, Mind Games* (Kalamazoo, MI: Medieval Institute Publications, 2017), 131–44.

22. George Abbot, *An exposition vpon the prophet Ionah. Contained in certaine sermons, preached in S. Maries Church in Oxford* (London: Richard Field, 1600), 591.

23. Abbot, *Ionah*, 591.

24. For more on early modern providence, see Alexandra Walsham, *Providence in Early Modern England* (Oxford: Oxford University Press, 1999); and Keith Thomas, *Religion and the Decline of Magic: Studies in Popular Beliefs in Sixteenth and Seventeenth Century England* (London: Penguin Books, 1991), 90–133.

25. William Gouge, *The Extent of Gods Providence, Set out in a Sermon, Preached in Black-Fryers Church, V. Nov. 1623* bound with *Gods Three Arrowes: Plagve, Famine, Sword, In three Treatises* (London: George Miller, 1631), 379–80.

26. Bernard Capp, *Astrology and the Popular Press: English Almanacs 1500–1800* (London: Faber and Faber, 1979), 16.

27. William Perkins, *Fovre Great Lyers, Striuing who shall win the siluer Whetstone* (London: Robert Walde-graue, 1585), _1r, A8v.

28. Each of the Tudor monarchs, from Henry VII to Elizabeth, patronized a favorite astrologer. Elizabeth famously consulted mathematician and astrologer John Dee to help set the date of her coronation. Capp, *Astrology and the Popular Press*, 19.

29. George Herbert, *A Priest to the Temple, Or, The Countrey Parson His Character, and Rule of Holy Life* (London: T. Moxey, 1652), 122–3.

30. Herbert, *Countrey Parson*, 124.

31. Walsham, *Providence*, 13.

32. For more on puritan ideas about chance, see D. R. Bellhouse, "Probability in the Sixteenth and Seventeenth Centuries: An Analysis of Puritan Casuistry," *International Statistical Review* 56, no. 1 (April 1988): 63–74, particularly pp. 67–8. For more on the notion of prayer book Protestants, see Judith Maltby, *Prayer Book and People in Elizabethan and Early Stuart England* (Cambridge: Cambridge University Press, 2000).

33. James Balmford, *A Short and Plaine Dialogve Concerning the Vnlawfulnes of playing at Cards, or Tables, or any other Game consisting in chance* (London: Richard Boile, 1593), A1r. The shortness of the anti-gaming tract, which was published as both a book and a broadside, combined with its blackletter type—the font most accessible to poor men and women—indicates that Balmford was attempting to reach the widest possible audience. Intriguingly, Balmford has only written the lines of the preacher—the voice of the dialogue through which he himself speaks—in blackletter, putting his opponent's words in the less-legible italic type throughout. The only time the professor gets to speak in blackletter is when he asks an all-important question: "Why not for pleasure?" Balmford, *Dialogve*, A6r. *Oxford Dictionary of National Biography*, s.v. "Balmford, James (b. c. 1556, d. after 1623), *Church of England clergyman*." For more on early modern fonts, see Keith Thomas, "The Meaning of Literacy in Early Modern England," in *The Written Word: Literacy in Transition, Wolfson College Lectures 1985*, ed. Gerd Baumann (Oxford: Clarendon Press, 1986).

34. Balmford, *Vnlawfulnes*, A4r-A5r.

35. Balmford, *Vnlawfulnes*, A8v.

36. Walsham, *Providence*, 140.

37. James I, *Declaration to His Subjects Concerning Lawfull Sports*, in David Cressy and Lori Anne Ferrell, *Religion and Society in Early Modern England: A Sourcebook*, 2nd ed. (New York: Routledge, 2005), 169–70.

38. Balmford was the pastor at Rotherhithe, a London suburb and dock town. Margo Todd, "Providence, Chance and the New Science in Early Stuart Cambridge," *Historical Journal* 29, no. 3 (September 1986): 698.

39. Gataker treats "chance" and "casualtie" as essentially synonymous: e.g., Gataker, *Lots*, 3, 12.

40. Gataker, *Lots*, 12.

41. Gataker, *Lots*, 2.

42. Gataker, *Lots*, 4–5.

43. Gataker, *Lots*, 15.

44. Gataker, *Lots*, 146.

45. Gataker, *Lots*, 116.

46. Among Christians, generally, divination was considered impious as only God was privileged to know the future. This, of course, did not always stop Christians from employing divination for purposes as varied as finding a thief or determining the wisdom of a future activity. Thomas, *Religion and the Decline of Magic*, 253–64.

47. Gataker, *Lots*, 35.

48. Gataker, *Lots*, 360.

49. James Balmford, *A Modest Reply to Certaine Answeres, which Mr. Gataker B.D. in his Treatise of the Nature, & vse of Lotts, giveth to Arguments in a Dialogue concerning the Vnlawfulnes of Games consisting in Chance* (London: William Jaggard, 1623), 26. Bellhouse, "Puritan Casuistry," 71.

50. Balmford, *Modest Reply*, 89–91.

51. Balmford, *Modest Reply*, 109.

52. The use of oaths in gaming was a common practice, compounding the offensiveness of the games themselves. See, e.g., Charles Cotton, *The Compleat Gamester, or, Instructions How to play at Billiards, Trucks, Bowls, and Chess Together with all manner of usual and most Gentile Games either on Cards or Dice* (London: A. M., 1674), 20; Balmford, *Modest Reply*, 87–8.

53. Gataker's temper also manifested in the conclusion of his treatise, where he lectured his opponent on the proper rules of debate. Thomas Gataker, *A Ivst Defence of Certaine Passages in a former Treatise concerning the Nature and Vse of Lots* (London: John Haviland, 1623), A1r, Mm4r.

54. James Balmford, *A Short Dialogve Concerning the Plagves Infection* (London: Richard Boyle, 1603), 67.

55. Gataker, *Ivst Defence*, 3–4.

56. Balmford, *Plagves Infection*, 45, 49.

57. Balmford, *Plagves Infection*, 67.

58. Todd, "Providence," 698–9. For more on the debates among Gataker, Balmford, and Ames, see Bellhouse, "Puritan Casuistry," 68–72.

59. John Downe, *A Defense of the Lawfulnesse of Lots in Gaming Against the Arguments of N.N.*, printed in *Certaine Treatises of the Late Reverend and Learned Devine, Mr Iohn Downe, Rector of the Church of Instow in Devonshire* (London: Iohn Lichfield, 1633). For more on reactions to Gataker's work, see Thomas, *Religion*, 143–5.

60. C. F. Trenerry, *The Origin and Early History of Insurance* (London: P. S. King, 1926).

61. James Franklin, *The Science of Conjecture: Evidence and Probability Before Pascal* (Baltimore: Johns Hopkins University Press, 2001), 274. Guido Rossi, *Insurance in Elizabethan England: The London Code* (Cambridge: Cambridge University Press, 2016), 50.

62. Rossi, *Insurance in Elizabethan England*, 41, 43.

63. Rossi, *Insurance in Elizabethan England*, 55.

64. David Jenkins and Takau Yoneyama, eds., *History of Insurance* (London: Pickering and Chatto, 2000), 1:xxix. Geoffrey Clark, *Betting on Lives: The Culture of Life Insurance in England, 1695–1775* (Manchester: Manchester University Press, 1999), 20.

65. 43 Eliz, c. 12. Keble, *Statutes at Large*, 952–3.

66. Rossi, *Insurance in Elizabethan England*, 429–30. Eighteenth-century British slave trade insurance policies were limited to the usual maritime hazards, such as shipwrecks and piracy, but excluded any sort of hazard that was specific to human cargo, such as insurrections or illness. These sorts of exclusions would eventually lead to the infamous Zong massacre. The British Parliament subsequently passed a series of laws specifically to forbid the practice of jettisoning enslaved people for insurance purposes. Clark, *Betting on Lives*, 17.

67. It was also possible to take out an insurance policy on rents or annuities. Rossi, *Insurance in Elizabethan England*, 411. Clark, *Betting on Lives*, 14–18.

68. These policies also had steep penalties—up to 20%—if they paid out on someone who was presumed dead and that person showed up alive again. Rossi, *Insurance in Elizabethan England*, 412, 416, 419–20.

69. Trenerry, *Early History*, 277–8.

70. There are only three known examples of sixteenth-century London life insurance policies. Rossi, *Insurance in Elizabethan England*, 414–5.

71. Miranda Kaufmann, *Black Tudors: The Untold Story* (London: Oneworld Publications, 2017), 15–16, 156–7.

72. For more on credit in early modern England, see Craig Muldrew, *The Economy of Obligation: The Culture of Credit and Social Relations in Early Modern England* (New York: St. Martin's Press, 1998).

73. William West, *The First Part of Simboleography* (London: Companie of Stationers, 1610), Qq5r-v.

74. Clark, *Betting on Lives*, 20.

75. Jacob F. Field, *London, Londoners and the Great Fire of 1666: Disaster and Recovery* (New York: Routledge, 2018), 19.

76. A. Newbold, *Londons Improvement, and the Builders Security, &c.* (London: Thomas Milbourn, 1680), reproduced in Jenkins and Yoneyama, *History of Insurance*, 1:5.

77. Newbold, *Londons Improvement*, 1:8.

78. Harry M. Johnson, "The History of British and American Fire Marks," *Journal of Risk and Insurance* 39, no. 3 (September 1972): 406. These insurance rates were advertised in the *Mercurius Civicus* on May 12, 1680. Jenkins and Yoneyama, *History of Insurance*, 1:14.

79. Anonymous, *At a Common Council holden in the chamber of the Guildhall. Proposals . . . for Insuring of Houses in Cases of Fire* (London, 1681), reproduced in Jenkins and Yoneyama, *History of Insurance*, 1:32–3.

80. Clark, *Betting on Lives*, 7.

81. Francis Boyer Relton, *An Account of the Fire Insurance Companies* (London: Swan Sonnenschein, 1893), 11–13. Brian Wright, *Insurance Fire Brigades 1680–1929: The Birth of the British Fire Service* (Chalford, England: Tempus, 2008), 19.

82. Johnson, "Fire Marks," 406.

83. A series of laws passed in 1707–8 provided rewards for the first three fire engines to show up at any fire, thus providing another possible income stream for the fire brigades. Some firemen had a further retainer from their companies, and all firemen were officially exempt from being impressed by the navy. Wright, *Fire Brigades*, 42–4, 73–4.

84. Approximately half of the early insurance companies were established on a mutual basis. The Sun Fire Office had approximately 130 shareholders. Robin Pearson, *Insuring the Industrial Revolution: Fire Insurance in Great Britain, 1700–1850* (Burlington, VT: Ashgate, 2004), 62–3. Wright, *Fire Brigades*, 38.

85. Johnson, "Fire Marks," 406. Jenkins and Yoneyama, *History of Insurance*, 1:xx.

86. Clark, *Betting on Lives*, 21, 33. Jenkins and Yoneyama, *History of Insurance*, 1:xxx.

87. Daniel Defoe, *Essays Upon Several Projects: Or, Effectual Ways for advancing the Interest of the Nation* (London: Thomas Ballard, 1702), 171–2.

88. Clark, *Betting on Lives*, 1, 45.

89. Jenkins and Yoneyama, *History of Insurance*, 1: xxv–xxvi.

90. Philip Ziegler, *The Black Death* (New York: Harper Torchbooks, 1969), 224–30. For more on the origins and geographical extent of the Second Plague Pandemic, see Monica H. Green, "The Four Black Deaths," *American Historical Review* 125, no. 5 (2020): 1601–1631; "Putting Africa on the Black Death Map: Narratives from Genetics and Gistory," *Afriques* 9 (2018), https://doi.org/10.4000/afriques.2125.

91. Extrapolating outward from a small sample to all of England, Paul Slack estimated that at least 658,000 people died of the plague during approximately the same time period, meaning that one in three English plague deaths occurred in London. Paul Slack, *The Impact of Plague in Tudor and Stuart England* (Oxford: Clarendon Press, 1990), 174.

92. Paul Slack, "Perceptions of Plague in Eighteenth-century Europe," *Economic History Review* (April 2021): 1–19, https://doi.org/10.1111/ehr.13080.

93. Slack, *Impact of Plague*, 208. Kristin Heitman, "Authority, Autonomy and the First London Bills of Mortality," *Centaurus* 62, no. 2 (2020): 278, https://doi.org/10.1111/1600-0498.12305. For more on late medieval and sixteenth-century European authorities' use of lists and numbers to assess mass mortality, see John Gagné, "Counting the Dead: Traditions of Enumeration and the Italian Wars," *Renaissance Quarterly* 67 (2014): 791–849.

94. Londoners from households where someone was sick were required to hang bundles of straw outside their windows for forty days and carry a white stick whenever they left home. Oxford implemented a similar policy in April. Slack, *Impact of Plague*, 201–2.

95. For the significance of the mental shift from quantification, see Mary Poovey, *A History of the Modern Fact: Problems of Knowledge in the Sciences of Wealth and Society* (Chicago: University of Chicago Press, 1998), especially xii, 4. This was

not the first time early modern Europeans had enumerated their dead with accuracy, e.g., there exists a fifteenth-century "inventory" from Paris's Hôtel-Dieu that includes monthly mortality figures for dead patients. Gagné, "Counting the Dead," 809. By 1551, the city of Geneva had also instituted its own quantitative bill of mortality.

96. SP 1/49 f.210 (Gale Document Number MC4301102356). These "plague" deaths may have been due to sweating sickness, which was ravaging the city in 1528, rather than plague caused by Y. pestis. The term "plague" was still used generally for many contagious diseases until the late sixteenth century. Only in the seventeenth century was it used to refer to just the disease caused by Y. pestis that we call plague today. Slack, *Impact of Plague*, 64–5.

97. Stephen Greenberg, "Plague, Printing, and Public Health," *Huntington Library Quarterly* 67, no. 4 (December 2004): 513, 516.

98. John Stow and Edmund Howes, *Annales, or A Generall Chronicle of England* (London: Richard Meighen, 1631), 656–7. William Ogle, "An Inquiry into the Trustworthiness of the Old Bills of Mortality," *Journal of the Royal Statistical Society* 55, no. 3 (September 1892): 437–9.

99. John Graunt, *Natural and Political Observations Mentioned in a following Index, and made upon the Bills of Mortality* (London: Thomas Roycroft, 1662), 33. John Graunt, *Reflections on the Weekly Bills of Mortality For the Cities of London and Westminster* (London: Samuel Speed, 1665), 1.

100. France started keeping its own baptism registers in 1539 and added weddings and funerals in 1579. The Council of Trent further required all Catholic parishes to begin keeping registers of christenings and marriages, while funerals were added by Pope Paul V in 1614. Daston, *Classical Probability*, 126. See also Chapter 7.

101. Slack, *Impact of Plague*, 111.

102. Slack, *Impact of Plague*, 128–9.

103. Cornelius Walford, "Early Bills of Mortality," *Transactions of the Royal Historical Society* 7 (1878): 234, 245.

104. Graunt, *Reflections*, 10. They continued to be published until the middle of the nineteenth century, by which point they had been replaced by the weekly returns of the secular General Registrar's Office.

105. Slack, *Impact of Plague*, 120.

106. Graunt, *Natural and Political Observations*, 4.

107. See, for example, England and Wales, *Whereas An Ordinance Was Lately Made By Both Houses of Parliament, For the Speedy Supply of the Cities of London and Westminster* (London, 1643), A1r.

108. Greenberg, "Plague," 510. Walford, "Early Bills," 216. Graunt, *Natural and Political Observations*, 11.

109. Raworth's sermon also demonstrates that people noted causes of death other than plague: "When I take a view of our weekly Bills of Mortality in *London*, I finde a report of so many dying in one Parish of a Fever, of so many dying in another Parish of a Consumption, &c." Francis Raworth, *Blessedness, or, God and the World Weighed in the Balances of the Sanctuary and the World found too light. Preached in a Sermon at Pauls* (London: T. Maxey, 1656), 7.

110. William Sanderson, *A Compleat History of the Life and Raigne of King Charles From His Cradle to his Grave* (London: Humphrey Moseley, Richard Tomlins, George Sawbridge, 1658), 897. These might have been manuscript bills rather than printed bills, in the expanded form usually sent weekly to the monarch, Lord Mayor of London, and archbishop of Canterbury. Craig Spence, *Accidents and Violent Death in Early Modern London: 1650–1750* (Woodbridge, Suffolk: Boydell Press, 2016), 53. At least three dozen of these expanded form bills survive, which show the parish-by-parish breakdown of all causes of death, not just plague.

111. Company of Parish Clerks, *LONDON'S Dreadful Visitation: Or, A COLLECTION of All the Bills of Mortality For this Present Year* (London: E. Cotes, 1665), A2v.

112. Greenberg, "Plague," 517–22. Slack, *Impact of Plague*, 151.

113. Greenberg, "Plague," 525.

114. I. D., *Salomons Pest-Hovse, or Tovver-Royall* (London: Henry Holland, 1636), A4v.

115. LMA, ACC/0262/043, f. 23r, 96r.

116. Historical *London Gazette* issues can be found through http://www.london-gazette.co.uk.

117. Slack, *Impact of Plague*, 239, 245.

118. While not a factual account of the 1665 epidemic, Defoe's account was written in response to the 1721–2 Marseilles outbreak that caused a major plague scare in England. Daniel Defoe, *A Journal of the Plague Year* (London: Falcon Press, 1950), 8–10, 18, 56, 111. Text taken from the first edition, London, 1722.

119. Samuel Pepys, *The Diary of Samuel Pepys*, ed. Robert Latham and William Matthews (Berkeley: University of California Press, 1972), 6:108.

120. Pepys, *Diary*, 6:180. Surviving bills indicate there were 2,010 plague deaths reported the prior week and 3,014 deaths overall. See, e.g., Company of Parish Clerks, *LONDON'S Dreadful Visitation*, I4v.

121. Pepys, *Diary*, 6:187.

122. Pepys, *Diary*, 7:376–7.

123. Balmford, *Plague's Infection*, A4v.

124. I. D., *Salomons Pest-Hovse*, A4v.

125. Matthew Mead, *Solomon's Prescription For the Removal of the Pestilence: or, The Discovery of the Plague of our Hearts, in order to the Healing of that in our Flesh* (London: 1665), 3, 81.

126. Oliver Cromwell, *A Declaration of His Highnes the Lord Protector For a Day of Publick Thanksgiving, With An Order of His Highnes Council in Scotland For the Government thereof, For a Day of Publick Thanksgiving in Scotland* (Edinburgh: Christopher Higgins, 1658), 5.

127. I. D., *Salomons Pest-Hovse*, A4v.

128. Pepys, *Diary*, 6:142, 145, 147.

129. Pepys, *Diary*, 6:163, 180.

130. Defoe, *Journal*, 15–8.

131. Slack, *Impact of Plague*, 256.

132. Pepys, *Diary*, 6:187.

133. For more on the searchers of the dead and debates on the accuracy of the bills of mortality, see Jeremy Boulton and Leonard Schwarz, "Yet Another Inquiry into the

Trustworthiness of Eighteenth-century London's Bills of Mortality," *Local Population Studies* 85 (2010): 28–45; Deborah Harkness, "A View from the Streets: Women and Medical Work in Elizabethan London," *Bulletin of the History of Medicine* 82, no. 1 (2008): 52–85; Wanda S. Henry, "Women Searchers of the Dead in Eighteenth- and Nineteenth-Century London," *Social History of Medicine* 29 no. 3 (2016): 445–66; William Ogle, "An Inquiry into the Trustworthiness of the Old Bills of Mortality," *Journal of the Royal Statistical Society* 55, no. 3 (1892): 437–60; Richelle Munkhoff, "Reckoning Death: Women Searchers and the Bills of Mortality in Early Modern London," in Jennifer C. Vaught, ed., *Rhetorics of Bodily Disease and Health in Medieval and Early Modern England* (London: Routledge, 2010); and Spence, *Accidents and Violent Death*, 44–50.

134. Graunt, *Reflections*, 2.
135. Spence, *Accidents and Violent Death*, 47–8.
136. Graunt, *Reflections*, 2.
137. Pepys, *Diary*, 6:289.

Chapter 6

1. Anne Whiteman, ed., *The Compton Census of 1676*, British Academy Records of Social and Economic History, n.s., 10 (London: Oxford University Press, 1986), xxiv.
2. For more on the government of early modern England, see Edward Higgs, *The Information State in England* (New York: Palgrave Macmillan, 2004), 30–39; Michael J. Braddick, *State Formation in Early Modern England c. 1550–1700* (Cambridge: Cambridge University Press, 2000); Steve Hindle, *The State and Social Change in Early Modern England c. 1550–1640* (New York: St. Martin's Press, 2000).
3. Whiteman, *Compton Census*, xxiv, xxxvi.
4. Whiteman, *Compton Census*, xlii–iii.
5. This has led some historians, notably Thomas Richards, to argue that the census results were manipulated, "driving up the ratio of Conformists to Dissenters as high as possible." However, Anne Whiteman's detailed examination of both the original returns and tabulated data indicates that most discrepancies were arithmetical mistakes, copying slips, or adjustments meant to compensate for inconsistencies in the collection process, such as the failure to include women. Whiteman, *Compton Census*, xxiv, xlviii–xlix. For a detailed analysis and rebuttal of Richards's arguments, see Appendix C, xcii–c.
6. Paul Slack, "Government and Information in Seventeenth-Century England," *Past and Present* 184 (August 2004): 46–7.
7. For more on quantitative arguments used by late seventeenth- and eighteenth-century political outsiders, see William Deringer, *Calculated Values: Finance, Politics, and the Quantitative Age* (Cambridge, MA: Harvard University Press, 2018).
8. Thomas Grantham, *A Marriage Sermon. A Sermon Called A Wife Mistaken* (London, 1641), 12.

9. Livy, *The Romane Historie Written by T. Livius of Padua . . . Translated out of Latine into English, by Philemon Holland* (London: Adam Islip, 1600), 655.

10. Arthur Jackson, *A Help For The Understanding of the Holy Scripture . . . Containing certain short notes of exposition upon the five books of Moses* (London: Roger Daniel, 1643), 321.

11. Jean Bodin, *The Six Bookes of A Common-Weale . . . Out of the French and Latine Copies, done into English by Richard Knolles* (London: G. Bishop, 1606), 639.

12. William Perkins, *A Commentarie, or Exposition, vpon the fiue first Chapters of the Epistle to the Galatians* (Cambridge: John Legat, 1604), 503.

13. Andrew Willet, *An Harmonie Vpon the Second Booke of Samvuel* (Cambridge: Cantrell Legge, 1614), 139.

14. Hugh Latimer, *The seconde Sermon of Maister Hughe Latimer, whych he preached before the Kynges maiestie, w[ith]in his graces Palayce at Westminster* (London: Ihon Day, 1549), S5v.

15. Willet, *Harmonie*, 139.

16. Anthony Burgess, *A Treatise of Original Sin* (London: 1658), 238. William Guild, *The Throne of David, or An Exposition of the Second of Samuell* (Oxford: W. Hall, 1659), 325.

17. James Howell, *Proedria Vasilike, A Discourse Concerning the Precedency of Kings* (London: James Cottrel, 1664), 84.

18. The fifteenth and tenth was a direct tax in which each parish and township was assigned a quota based on the amount of tax they had paid in 1334, minus poverty deductions. The parish or town leadership was then allowed to decide among themselves how they wanted to distribute their quota among the population. R. W. Hoyle, "Crown, Parliament and Taxation in Sixteenth-Century England," *English Historical Review* 109, no. 434 (November 1994): 1175.

19. Paul Griffiths, "Surveying the People," in Keith Wrightson, ed., *A Social History of England: 1500–1750* (Cambridge: Cambridge University Press, 2017), 40. Paul Slack, "Counting People in Early Modern England: Registers, Registrars, and Political Arithmetic," *English Historical Review* 137, no. 587 (August 2022): 1120–1. For more on Henry VIII's royal ministers, see G. R. Elton, *The Tudor Revolution in Government: A Study of Administrative Changes in the Reign of Henry VIII* (Cambridge: Cambridge University Press, 1969).

20. Hoyle, "Crown," 1178.

21. Ian W. Archer, "The Burden of Taxation on Sixteenth-Century London," *Historical Journal* 44, no. 3 (September 2001): 615. See, for example, the Hertfordshire muster books, which include several lists of those required to pay toward the muster, supply equipment, or maintain a watch on the local beacon. Ann J. King, ed., *Muster Books for North and East Hertfordshire, 1580–1605*, Hertfordshire Record Publications 12 (Hitchin: Hertfordshire Record Society, 1996), 5, 146–7, 170, 175–207.

22. Slack, "Government," 38.

23. Thomas Elyot, *The Dictionary of syr Thomas Eliot knyght* (London: Thomae Bertheleti, 1538), C5v. This association continued throughout the early modern period with

other authors variously defining a census as the "valuation of euery mans goods," "poll mony," "tribute or taxe," and "*revenues*, which a man is esteemed to have, and accordingly is rated and pays Subsidies." Bodin, *Six Bookes*, 637. Andrew Willet, *Hexapla in Genesin & Exodum: That is, A sixfold commentary upon the two first Bookes of Moses, being Genesis and Exodvs* (London: John Haviland, 1633), 587. William Howell, *An Institution of General History from the beginning of the World to the Monarchy of Constantine the Great* (London: Henry Herringman, 1661), 167. Simon Patrick, *A Paraphrase Upon the Books of Ecclesiastes and the Song of Solomon with arguments to each chapter and annotations thereupon* (London: W.H., 1700), 81.

24. Hoyle, "Crown," 1175-1177. Archer, "Burden," 605.

25. Corporation of London, *Commyssioners for our soueraygne lorde kynge* (London: 1515), A1r.

26. Archer, "Burden," 601.

27. Archer, "Burden," 610.

28. Braddick, *State Formation*, 237-8.

29. Parliamentary subsidies totaled £105,643 in the 1540s and £77,883 in the 1550s. After the oath was dropped, revenues dramatically decreased to £33,417 in the 1560s and £27,821 in the 1570s. Even with near-yearly subsidies, they only rose again to £38,607 in the 1580s and £57,383 in the 1590s. Archer, "Burden," 608-9.

30. Braddick, *State Formation*, 254.

31. Beginning in 1663, the hearth tax included lists of both those "chargeable" and exempt from the tax. From 1684 onward, assessors were also required to include a list of household residents and the names of those responsible for empty houses. Paul Seaman, ed., *Norfolk Hearth Tax Exemption Certificates 1670-1674: Norwich, Great Yarmouth, King's Lynn and Thetford*, Norfolk Record Society 65 (London: British Record Society, 2001), xxvi, xxxiii-iv. Paul Slack, "Measuring the National Wealth in Seventeenth-Century England," *Economic History Review*, n.s., 57, no. 4 (November 2004): 612.

32. Braddick, *State Formation*, 256. The legacy of this tax can still be seen today in the bricked-up windows of historic houses.

33. James H. Cassedy, *Demography in Early America: Beginnings of the Statistical Mind, 1600-1800* (Cambridge, MA: Harvard University Press, 1969), 18-19. Robert V. Wells, ed., *The Population of the British Colonies in America Before 1776: A Survey of Census Data* (Princeton, NJ: Princeton University Press, 1979), 5, 7, 13. For a complete list of the 124 censuses conducted between 1623 and 1775, see pp. 8-11.

34. England began conducting regular censuses of its entire population only in 1801. Wells, *Population*, 7. Slack, "Government," 47.

35. Slack, "Government," 50.

36. Slack, "Counting People," 7.

37. Peter Burke, *A Social History of Knowledge from Gutenberg to Diderot* (Cambridge: Polity, 2000), especially pp. 117-19. Paul Griffiths, "Local Arithmetic: Information Cultures in Early Modern England," in Steve Hindle, Alexandra Shepard, and John Walter, eds., *Remaking English Society: Social Relations and Social Change in Early Modern England* (Woodbridge, Suffolk: Boydell Press,

2013), 116. Griffiths, "Surveying," 39. Paul M. Dover, *The Information Revolution in Early Modern Europe* (Cambridge: Cambridge University Press, 2021), 91.

38. Griffiths, "Local Arithmetic," 115. Griffiths, "Surveying," 47.

39. Griffiths, "Local Arithmetic," 128–9.

40. Slack, "Counting People," 4. For parallel efforts to track and tax immigrants, see W. Mark Ormord, Bart Lambert, and Jonathan Mackman, *Immigrant England, 1300–1550* (Manchester: Manchester University Press, 2019).

41. Slack, "Counting People," 5, 12.

42. The 1695 act required a payment of 2s. for a birth, 2s.6p. for a marriage, and 4s. for a burial. There were additional payments required of the rich, gentlemen, bachelors, and widowers, while those receiving poor relief were exempt. Colin Brooks, "Projecting, Political Arithmetic and the Act of 1695," *English Historical Review* 97, no. 382 (January 1982): 32.

43. Slack, "Counting People," 8.

44. Griffiths, "Local Arithmetic," 119–20.

45. John F. Pound, ed., *The Norwich Census of the Poor, 1570*, Norfolk Record Society 40 ([Norfolk]: Norfolk Record Society, 1971), 8–9. Slack, "Government," 40n24. Griffiths, "Surveying," 44.

46. Griffiths, "Surveying," 44–5.

47. Slack, "Counting People," 3–4. Paul Slack, *The Invention of Improvement: Information and Material Progress in Seventeenth-Century England* (Oxford: Oxford University Press, 2015), 44–45.

48. SP 1/49 f.210 (Gale Document Number MC4301102356). Stephen Greenberg, "Plague, Printing, and Public Health," *Huntington Library Quarterly* 67, no. 4 (December 2004): 513. See also Chapter 6.

49. Slack, "Counting People," 8–9.

50. Paul Slack, *The Impact of Plague in Tudor and Stuart England* (Oxford: Clarendon Press, 1990), 111, 113, 120, 239.

51. I. D., *Salomons Pest-Hovse, or Tovver-Royall* (London: Henry Holland, 1636), 8–9.

52. Slack, "Counting People," 9. Gerard Malynes, *Englands View, in the Vnmasking of Two Paradoxes* (London: Richard Field, 1603), A2v–A3r, 6, 135–6.

53. Slack, "Government," 42.

54. *OED*, s.v. "population, *n.1.*" For an extensive discussion on the development of the population as a unit of analysis, see Ted McCormick, *Human Empire: Mobility and Demographic Thought in the British Atlantic World, 1500–1800* (Cambridge: Cambridge University Press, 2022).

55. Slack, "Government," 66.

56. Slack, "Counting People," 9.

57. Anonymous, *The Humble Petition and Remonstrance of Some Hundreds of Retaylers . . . For the Restoring of Farthing Token* (London, 1644), A3v.

58. Rice Bush, *The Poor Mans Friend, or A Narrative of what progresse many worthy Citi- [sic] of London have made in that Godly Work of providing for the Poor. With An Ordinance of Parliament for the better carrying on of the Work* (London: A. M., 1649), 18.

59. James Howell, *Londinopolis; An Historicall Discourse or Perlustration Of the City of London* (London: J. Streater, 1657), 389.
60. Richard Whitlock, *Zootomia, or Observations on the Present Manners of the English* (London: Tho. Roycroft, 1654), 60. Thomas Culpeper, *Morall Discourses and Essayes, Upon Severall Select Subjects* (London: S. G. for Charles Adam, 1655), 281.
61. Marchamont Nedham, *Medela Medicinae. A Plea For the Free Profession, and a Renovation of the Art of Physick, Out of the Noblest and most Authentick Writers* (London: Richard Lownds, 1665), A1v-A2r, 1, 29.
62. Nedham, *Medela Medicinae*, 42-3.
63. Nedham, *Medela Medicinae*, A8r, 44-55.
64. John Graunt, *Natural and Political Observations Mentioned in a following Index, and made upon the Bills of Mortality* (London: Thomas Roycroft, 1662), 2-3.
65. Graunt, *Natural and Political Observations*, a1r-a4v.
66. Graunt, *Natural and Political Observations*, 71.
67. Graunt, *Natural and Political Observations*, 72-3.
68. Graunt, *Natural and Political Observations*, 74.
69. There were two editions of Graunt's *Natural and Political Observations* published in 1662, two further editions published in 1665, and a fifth edition published in 1676.
70. Matthew Hale, *The Primitive Origination of Mankind, Considered and Examined According to the Light of Nature* (London: William Godbid, 1677), 232.
71. *OED*, s.v. "political, *adj.* and *n.*" William Petty, *Political Arithmetick, or A Discourse Concerning The Extent and Value of Lands, People, Buildings . . .* (London: Robert Clavel, 1690), A2v-A3r. *Oxford Dictionary of National Biography*, s.v. "Petty, Sir William (1623-1687)," by Toby Barnard, http://oxforddnb.com/view/article/22069. For an extensive treatment of Petty and political arithmetic, see Ted McCormick, *William Petty and the Ambitions of Political Arithmetic* (Oxford: Oxford University Press, 2009).
72. Petty repeatedly used the biblical language of number, weight, and measure to justify his methods. William Petty, *England's Guide to Industry: or, Improvement of Trade for the good of all People in general* (London: R. Holt, 1683) bound with Edward Chamberlayne, *The Present State of England* (London: William Whitwood, 1683), Aa4v-Aa5r.
73. Graunt compared London mortality statistics with those of Romsey, the country parish where Petty was born. Slack, "Government," 64. McCormick, *William Petty*, 119, 131.
74. McCormick, *William Petty*, 119. Slack, "Counting People," 13.
75. McCormick, *William Petty*, 10, 206-7.
76. Richard Orpen, *An Exact Relation of the Persecutions, Robberies, and Losses, sustained by the Protestants of Killmare, in Ireland* (London: Thomas Bennet, 1689), 2.
77. Slack, "Government," 47n46.
78. McCormick, *William Petty*, 193-4.
79. That Petty could so quickly alter his earlier, anti-Catholic propositions to appeal to the first English Catholic monarch in over a century demonstrates how Petty's methods

were intended to serve the goals of the monarchy—and his own goal of monarchical preferment—rather than a personal political agenda. He was, after all, a man who landed on his feet after the Restoration despite his close political alliance and friendship with Lord Protector Oliver Cromwell's son Henry. McCormick, *William Petty,* 11, 258.

80. McCormick, *William Petty,* 131.
81. McCormick, *William Petty,* 217–8.
82. McCormick, *William Petty,* 223.
83. BL, MS Additional 72865, ff. 7–8 as cited in McCormick, *William Petty,* 178. For example, Graunt similarly was forced to make some guesses in his demographic analyses, and people from all over Europe subsequently reused his data for their own purposes. Ian Hacking, *The Emergence of Probability: A Philosophical Study of Early Ideas About Probability Induction and Statistical Inference,* 2nd ed. (Cambridge: Cambridge University Press, 2006), 128–9, 133.
84. Petty, *Political Arithmetick,* A4v.
85. William Petty, *Further Observation Upon the Dublin-Bills: Or, Accompts of the Houses, Hearths, Baptisms, And Burials in that City* (London: Mark Pardoe, 1686), 2–4.
86. Petty, *Dublin-Bills,* 6. McCormick, *William Petty,* 146.
87. Charles Davenant, *Discourses On the Public Revenues, and On the Trade of England* (London: James Knapton, 1698), 2. William Petty, "An Extract of two Essays in Political Arithmetick concerning the comparative Magnitudes &c. of London and Paris by Sir William Petty Knight, R.S.S.," *Philosophical Transactions of the Royal Society* 16 (1686), 152, as cited in Peter Buck, "Seventeenth-Century Political Arithmetic: Civil Strife and Vital Statistics," *Isis* 68, no. 1 (March 1977): 81.
88. McCormick, *William Petty,* 287–8.
89. John Houghton, *A Proposal for the Improvement of Husbandry and Trade* (London, 1691), 1.
90. Gregory King, *Natural and Political Observations on the State of England* (London, 1802) as cited in McCormick, *William Petty,* 294.
91. For an analysis of how King came by one of his population estimates, see D. V. Glass, "Gregory King's Estimate of the Population of England and Wales, 1695" *Population Studies* 3, no. 4 (March 1950): 338–374. Slack, "Measuring," 623–4. McCormick, *William Petty,* 294.
92. Davenant, *Discourses,* 2.
93. Davenant, *Discourses,* 13–4.
94. Charles Davenant, *An Essay Upon the Ways and Means of Supplying the War* (London: Jacob Tonson, 1695), 134–5.
95. McCormick, *William Petty,* 299. Like the financial explorations of the 1690s Commission of Public Accounts or the public debates over the 1706 calculation of the Equivalent, the quantitative analyses of political arithmetic were particularly useful to relative outsiders attempting to comprehend, influence, or even attack established loci of political power. Deringer, *Calculated Values,* especially pages 9–12, 23–4. For more on eighteenth-century political arithmetic, see also Peter Buck, "People

Who Counted: Political Arithmetic in the Eighteenth Century," *Isis* 73, no. 1 (March 1982): 28-45; Julian Hoppit, "Political Arithmetic in Eighteenth-Century England," *Economic History Review* 49, no. 3 (1996): 516-540; McCormick, *Human Empire.*

Epilogue

1. Thomas Dekker, *The Pleasant Comedie of Old Fortunatus* (London: S. S. for William Aspley, 1600), L3v.
2. Cathy O'Neill, *Weapons of Math Destruction: How Big Data Increases Inequality and Threatens Democracy* (New York: Crown, 2016). Safiya Umoja Noble, *Algorithms of Oppression: How Search Engines Reinforce Racism* (New York: New York University Press, 2018). Jacqueline Wernimont, *Numbered Lives: Life and Death in Quantum Media* (Cambridge, MA: MIT Press, 2019).

Bibliography

Manuscripts

British Library
MS Additional 4239 Adam Martindale's Autobiography
MS Additional 4473 William Senior Mathematical Advertisement, 1641
MS Additional 27999–28005 Oxinden Correspondence
MS Additional 45513 An Easy Way To the Art of Numbers by Michael Wicks, 1658
MS Arundel 293 Wilboldt Imhoff of Nürmberg's Cyphering Book
MS Lansdowne 98 Scheme of an Academy for Wards
MS Lansdowne 213 Relation of a Short Survey of 26 Counties
MS Lansdowne 1181 Keils Arithmetick
MS Sloane 1428 A Book of Arithmetick

Bodleian Library, Oxford University
MS Aubrey 10 An Idea of Education of Young Gentlemen

Essex Records Office
D/DHe/E8 Fire Insurance Certificates
D/P 94/5/1 Chelmsford Account Book, 1557–1668
D/Q 6/1/1 Papers of the Free Grammar School in Earles Colne

Guildhall Library
MS 25491 St. Paul's Cathedral Account Book, 1678 Rebuilding
Store 424.9 Bills of Mortality, 1602–66

Hatfield House Archives
CP 30/80 The Debts of the Late Bishop of Winchester
CP 109/79 Purveyance

Huntington Library
MS HA Inventories Inventory of Earl of Huntingdon's Goods, 1596
MS HA School Exercises Mathematical Commonplace Book, via John Speidell 1623

London Metropolitan Archives
E/PYM/624–31 London Fire Insurance Policies, 1682–9
ACC/0262/043 Letters from Edward Wood to John Pack, 1665
WJ/SPD1–71 Statements of Loss of Goods to Fire, 1672 and 1697

The National Archives of the United Kingdom

C171/45 Account Book of Duke of Richmond and Lennox, 1664–5
LC2/1 State Funeral Accounts
SP44/245 Insurance Schemes
SP46/157 Exchequer Accounts, 1667–82
SP105/109 Charles II On Founding the Royal Mathematical School
E101/274/9 Account Book of Robert Recorde, Surveyor of the Mines
E179/251/9 Hearth Tax Papers
E344/1 Valor Ecclesiasticus

Early Modern Printed Works

Abbot, George. *An exposition vpon the prophet Ionah. Contained in certaine sermons, preached in S. Maries Church in Oxford*. London: Richard Field, 1600.

Allen, Charles. *The Battailes of Crescey, and Poictiers vnder the leading of King Edward the Third of that name; And his Sonne Edward Prince of Wales, named the Blacke*. London: Thomas Purfoot, 1631.

Allestree, Richard. *Allestree. 1621. A New Almanack and Prognostication*. London: Company of Stationers, 1621.

Allestree, Richard. *Allestree. 1641. A New Almanack and Prognostication*. London: T. Coates, 1641.

Alleyn, Henry. *Alleyn 1607. A double Almanacke & Prognostication*. London: Company of Stationers, 1607.

Anonymous. *1679. A Yea and Nay Almanack For the People Call'd By the Men of the World Quakers*. London, 1679.

Anonymous. *An Account of the Solemn Reception of Sir Iohn Robinson, Lord-Maior at St. Pauls Cathedral, the day of his inauguration*. London: Joshua Coniers, 1662.

Anonymous. *The ABC with The Catechisme: That is to say, an Instruction to bee taught and learned of euery Child, before he be brought to be confirmed by the Bishop*. London: n.p., 1633.

Anonymous. *The arte and science of arismetique*. London: Rychard Fakes, 1526.

Anonymous. *An introduction for to lerne to recken with the pen or with the counters*. London: Nycolas Bourman, 1539.

Anonymous. *An introduction for to lerne to recken with the pen, or with the counters accordyng to the trewe cast of Algorisme, in hole numbers or in broken*. London: Ihon Herford, 1546.

Anonymous. *An introduction for to lerne to recken with the pen, or wyth the counters accordynge to the trew cast of algorisme, in hole numbers or in broken*. London: S. Mierdman for John Waley, 1552.

Anonymous. *An introduction for to learne to recken wyth the pen, or with the counters, according to the true rule of algorisme, in whole numbers or in broken*. London: John Awdley, 1566.

Anonymous. *An introduction of algorisme, to learn to reckon with pen or with the counters, in whole numbers or in broken*. London: John Awdeley, 1574.

Anonymous. *An introduction of algorisme, to learne to reckon with the penne, or with the counters, in whole numbers or in broken*. London: John Charlewood, 1581.

Anonymous. *At a Common Council holden in the chamber of the Guildhall. Proposals . . . for Insuring of Houses in Cases of Fire*. London, 1681.

Anonymous. *The Humble Petition and Remonstrance of Some Hundreds of Retaylers . . . For the Restoring of Farthing Token*. London, 1644.

Anonymous. *Lord Have Mercy Upon Us*. London: R. Young and M. Flesher, 1636.

Anonymous. *Observations on the Proposals of the City to Insure Houses in case of Fire*. London, 1681.

Arbuthnot, John. *An essay on the usefulness of mathematical learning, in a letter from a gentleman in the city to his friend in Oxford*. Oxford: Anth. Peisley, 1701.

Ascham, Roger. *The Scholemaster*. London: John Daye, 1570.

Astell, Mary. *An Essay in Defence of the Female Sex. In which are inserted the Characters of a Pedant, A Squire, A Beau, A Vertuoso, A Poetaster, A City-Critick, &c*. London: A Roper, E. Wilkinson and R. Clavel, 1696.

B., R. *An Idea of Arithmetick*. London: J. Flesher, 1655.

Babington, Gervase. *Certaine plaine, briefe, and comfortable notes vpon euerie chapter of Genesis*. London: A. Jeffes and P. Short, 1592.

Baker, Humfrey. *The Well Sprynge of Sciences*. London: Rouland Hall, 1562.

Baker, Humfrey. *The welspring of sciences*. London: Henry Denham, 1564.

Baker, Humfrey. *The vvell spryng of sciences*. London: Ihon Kyngston, 1568.

Baker, Humfrey. *The well springe of sciences*. London: Thomas Purfoote, 1574.

Baker, Humfrey. *The well spring of sciences*. London: Thomas Purfoot, 1576.

Baker, Humfrey. *The well spring of sciences*. London: Thomas Purfoot, 1580.

Baker, Humfrey. *The well spring of sciences*. London: Thomas Purfoote, 1583.

Baker, Humfrey. *The wel spring of sciences*. London: Thomas Purfoote, 1591.

Baker, Humfrey. *The well spring of sciences*. London: Thomas Purfoot, 1598.

Baker, Humfrey. *The well spring of sciences*. London: Thomas Purfoot, 1602.

Baker, Humfrey. *The well-spring of sciences*. London: Thomas Purfoot, 1607.

Baker, Humfrey. *The well-spring of sciences*. London: Thomas Purfoot, 1612.

Baker, Humfrey. *The vvell-spring of sciences*. London: Thomas Purfoot, 1617.

Baker, Humfrey. *The well-spring of sciences*. London: Thomas Purfoot, 1631.

Baker, Humfrey. *The well-spring of sciences*. London: M.F., 1646.

Baker, Humfrey. *The well-spring of sciences*. London: J. Flesher, 1650.

Baker, Humfrey. *The well-spring of sciences*. London: R. & W. L., 1655.

Baker, Humfrey. *The wel-spring of sciences*. London: Andrew Kemb, 1659.

Baker, Humfrey. *Baker's Arithmetick*. Edited by Henry Phillippes. London: E. C. & A. C., 1670.

Balmford, James. *A Modest Reply to Certaine Answeres, which Mr. Gataker B.D. in his Treatise of the Nature, & vse of Lotts, giveth to Arguments in a Dialogue concerning the Vnlawfulness of Games consisting in Chance*. London: William Jaggard, 1623.

Balmford, James. *A Short Dialogve Concerning the Plagves Infection*. London: Richard Boyle, 1603.

Balmford, James. *A Short and Plaine Dialogve Concerning the Vnlawfulnes of playing at Cards, or Tables, or any other Game consisting in chance*. London: Richard Boile, 1593.

Barton, William. *Arithmeticke abreviated. Teaching the art of tennes or decimals to worke all questions in fractions as whole numbers, without reduction*. London: John Beale, 1634.

Baxter, Richard. *The Right Method for a Settled Peace of Conscience and Spiritual Comfort in 32 Directions*. London: T. Underhill, F. Tyton, and W. Raybould, 1653.

Best, George. *A Trve Discovrse of the late voyages of discouerie, for the finding of a passage to Cathaya, by the Northvveast, vnder the conduct of Martin Frobisher Generall*. London: Henry Bynnyman, 1578.

B., R.. *An Idea of Arithmetick at First Designed for the Use of the Free-Schoole at Thurlow in Suffolk*. London: J. Flesher, 1655.

Blundevill, Thomas. *M. Blundeville his exercises, containing sixe treatises*. London: Iohn Windet, 1594.

Blundevill, Thomas. *M. Blundeuile his exercises, containing eight treatises*. London: Iohn Windet, 1597.

Blundevill, Thomas. *M. Blundeuile his exercises, containing eight treatises*. London: Iohn Windet, 1606.

Blundevill, Thomas. *M. Blundeuile his exercises, containing eight treatises*. London: William Stansby, 1613.

Blundevill, Thomas. *M. Blundeuile his exercises, contayning eight treatises*. London: William Stansby, 1622.

Blundevill, Thomas. *Mr. Blundevil his exercises, contayning eight treatises*. London: Richard Bishop, 1636.

Blundevill, Thomas. *Mr. Blundevil his exercises, contayning eight treatises*. London: Richard Bishop, 1638.

Bodin, Jean. *The Six Bookes of A Common-Weale . . . Out of the French and Latine Copies, done into English by Richard Knolles*. London: G. Bishop, 1606.

Bretnor, Thomas. *Bretnor 1609 a new almanacke and prognostication*. London: Company of Stationers, 1609.

Bretnor, Thomas. *Bretnor. 1616*. London: Company of Stationers, 1616.

Bridges, Noah. *Lux Mercatoria Arithmetick natural and decimal*. London: R. I., 1661.

Brinsley, John. *Lvdvs Literarivs: or, the Grammar Schoole: shewing how to proceede from the first entrance into learning, to the highest perfection required in the Grammar Schooles*. London: Felix Kyngston, 1627.

Brinsley, John. *Calendar-Reformation*. London: Francis Neile, 1648.

Brown, George. *A Compendious, but a Compleat System Of Decimal Arithmetick. Containing more Exact Rules for ordering Infinities, than any hitherto extant*. Edinburgh: George Brown, 1701.

Browne, Daniel. *Browne. 1625. A New Almanack and Prognostication*. London: Company of Stationers, 1625.

Bulwer, John. *Chirologia, or, The naturall language of the hand composed of the speaking motions, and discoursing gestures thereof: whereunto is added Chironomia, or, The art of manuall rhetoricke*. London: Thomas Harper, 1644.

Burgess, Anthony. *A Treatise of Original Sin*. London, 1658.

Bush, Rice. *The Poor Mans Friend, or A Narrative of what progresse many worthy Citi- [sic] of London have made in that Godly Work of providing for the Poor. With An Ordinance of Parliament for the better carrying on of the Work*. London: A. M., 1649.

Castlemaine, *An Account Of the Present War Between the Venetians & Turk; with the State of Candie*. London: J. M., 1666.

Cavendish, Margaret. *Natures picture drawn by fancies pencil to the life*. London: A. Maxwell, 1671.

Chamberlayne, Edward. *Angliae Notitia, or The Present State of England*. London: T. N. for John Martyn, 1669.

Chamberlayne, Edward. *The Second Part of the Present State of England Together With Divers Reflections Upon the Antient State Thereof*. London: Printed by T. N. for J. Martyn, 1671.

Chamberlen, Hugh. *Papers Relating to a Bank of Credit upon Land Security Proposed to the Parliament of Scotland*. Edinburgh: s.n., 1693.

Clavell, Robert. *A Catalogue of All the Books Printed in England since the Dreadful Fire of London in 1666, to the End of Michaelmas Term, 1672*. London: S. Simmons for R. Clavel, 1673.

Clement, Francis. *The Petie Schole with an English Orthographie*. London: Thomas Vautrollier, 1587.

Cocker, Edward. *The Tutor to Writing and Arithmetick*. London: Thomas Rooks, 1664.

Cocker, Edward. *Cockers Arithmetick*. Edited by John Hawkins. London: Thomas Passinger, 1678.

Cocker, Edward. *Cocker's Arithmetick*. Edited by John Hawkins. London: Thomas Passinger, 1680.

Cocker, Edward. *Cockers Arithmetick*. Edited by John Hawkins. London: Thomas Passinger, 1681.

Cocker, Edward. *Cocker's Arithmetick*. Edited by John Hawkins. London: Thomas Passinger, 1685.

Cocker, Edward. *Cocker's Arithmetick*. Edited by John Hawkins. London: Ralph Holt for Thomas Passinger, 1688.

Cocker, Edward. *Cocker's Arithmetick*. Edited by John Hawkins. London: Thomas Passinger, 1691.

Cocker, Edward. *Cocker's Arithmetick*. Edited by John Hawkins. London: J. R. for Thomas Passinger, 1694.

Cocker, Edward. *Cocker's Arithmetick*. Edited by John Hawkins. London: J. R. for Eben Tracey, 1697.

Cocker, Edward. *Cocker's Arithmetick*. Edited by John Hawkins. London: J. R. for Eben Tracey, 1698.

Cocker, Edward. *Cocker's Arithmetick*. Edited by John Hawkins. London: J. R. for Eben Tracey, 1700.

Cocker, Edward. *Cocker's Decimal Arithmetick*. Edited by John Hawkins. London: Thomas Passinger, 1685.

Cocker, Edward. *Cocker's Decimal Arithmetick*. Edited by John Hawkins. London: James Orme, 1695.

Coley, Henry. *Nuncius Coelestis: or Urania's Messenger*. London: J. Grover, 1679.

Company of Parish Clerks. *LONDON'S Dreadful Visitation: Or, A COLLECTION of All the Bills of Mortality For this Present Year*. London: E. Cotes, 1665.

Coote, Edmund. *The English School-Master*. London: B. Alsop, 1651.

Corporation of London. *Commyssioners for our soueraygne lorde kynge*. London, 1515.

Cotton, Charles. *The Compleat Gamester, or, Instructions How to play at Billiards, Trucks, Bowls, and Chess Together with all manner of usual and most Gentile Games either on Cards or Dice*. London: A. M., 1674.

Cowell, John. *The Interpreter: Or Booke Containing the Signification of Words*. Cambridge: Iohn Legate, 1607.

Cowell, John. *The Institutes of the Lawes of England Digested Into the Method of the Civill or Imperiall Institutions*. London: Tho. Roycroft for Jo. Ridley, 1651.

Cromwell, Oliver. *A Declaration of His Highnes the Lord Protector For a Day of Publick Thanksgiving, With An Order of His Highnes Council in Scotland For the Government thereof, For a Day of Publick Thanksgiving in Scotland*. Edinburgh: Christopher Higgins, 1658.

Cromwell, Oliver. *An Ordinance For bringing the Publique Revenues Of this Common-wealth into one Treasvry*. London: William du-Gard and Henry Hills, 1654.

Culpeper, Thomas. *Morall Discourses and Essayes, Upon Severall Select Subjects*. London: S. G. for Charles Adam, 1655.

D., I. *Salomons Pest-Hovse, or Tovver-Royall*. London: Henry Holland, 1636.

Davenant, Charles. *Discourses On the Public Revenues, and On the Trade of England*. London: James Knapton, 1698.

Davenant, Charles. *An Essay Upon the Ways and Means of Supplying the War*. London: Jacob Tonson, 1695.

Defoe, Daniel. *Essays Upon Several Projects: Or, Effectual Ways for advancing the Interest of the Nation*. London: Thomas Ballard, 1702.

Dekker, Thomas. *The Pleasant Comedie of Old Fortunatus*. London: S. S. for William Aspley, 1600.

Digges, Leonard. *An arithmeticall militare treatise, named Stratioticos: compendiously teaching the science of numbers, as vvell in fractions as integers*. Edited by Thomas Digges. London: Henrie Bynneman, 1579.

Digges, Leonard. *An arithmeticall vvarlike treatise named Stratioticos compendiously teaching the science of nombers as well in fractions as integers*. Edited by Thomas Digges. London: Richard Field, 1590.

Downe, John. *Certaine Treatises of the Late Reverend and Learned Devine, Mr Iohn Downe, Rector of the Church of Instow in Devonshire*. London: Iohn Lichfield, 1633.

Du Moulin, Pierre. *The Monk's Hood Pull'd Off, or, The Capvcin Fryar Described in Two Parts*. London, 1671.

Dury, John. *The Reformed School . . . Whereunto is Added I. An Idea of Mathematicks*. London: William Du-Gard, 1651.

Elestone, Sarah. *The Last Speech and Confession of Sarah Elestone at the Place of Execution Who Was Burned for Killing Her Husband, April 24. 1678*. London: T. D., 1678.

Elyot, Thomas. *The Dictionary of syr Thomas Eliot knyght*. London: Thomas Bertheleti, 1538.

England and Wales. *Whereas An Ordinance Was Lately Made By Both Houses of Parliament, For the Speedy Supply of the Cities of London and Westminster*. London, 1643.

Featley, Daniel. *Threnoikos, The House of Mourning*. London: G. Dawson, 1660.

Field, Theophilus. *A Christians Preparation to the Worthy Receiuing of the Blessed Sacrament of the Lords Supper*. London: Augustine Mathewes for Tho. Thorp, 1622.

Fine, Oronti. *Arithmetica Practica, Libris Qvatuor absoluta*. Paris: Simonis Colinaei, 1542.

Fitz-herbert, Anthony. *The New Natura Brevium, Of the most Reverend Judge Mr. Anthony Fitz-Herbert*. London: W. Lee, M. Walbank, D. Pakeman, and G. Bedett, 1652.

Foster, Samuel. *Of the Planetary Instruments: To what end they serve, and how they are to be used*. London, 1659.

Foxe, John. *Actes and Monuments of matters most speciall and memorable, happenyng in the Church, with an Vniuersall history of the same*. London: Iohn Daye, 1583.

Gadbury, John. *Ephemeris: or, A Diary Astronomical, Astrological, Meteorological, for the Year of Grace 1671*. London: J. C., 1671.

Gadbury, John. *Ephemeris: or, a Diary Astronomical & Astrological*. London: J. D., 1677.

Garcilasso de la Vega, Inca. *The Royal Commentaries of Peru in Two Parts*. London: Miles Flesher for Samuel Heyrick, 1688.

Gataker, Thomas. *A Ivst Defence of Certaine Passages in a former Treatise concerning the Nature and Vse of Lots*. London: John Haviland, 1623.

Gataker, Thomas. *Of the Natvre and Vse of Lots: A Treatise Historicall and Theologicall.* London: Edward Griffin, 1619.

Glanvil, Joseph. *Some Discourses, Sermons, and Remains of the Reverend Mr. Jos. Glanvil.* London: Anthony Honreck, 1681.

Goad, John. *Astro-meteorologica, or, Aphorisms and Discourses of the bodies coelestial.* London, 1686.

Godfrey, Michael. *A Short Account of the Bank of England.* London: John Whitlock, 1695.

Gouge, William. *The Extent of Gods Providence, Set out in a Sermon, Preached in Black-Fryers Church, V. Nov. 1623.* In *Gods Three Arrowes: Plagve, Famine, Sword, In three Treatises.* London: George Miller, 1631.

Grantham, Thomas. *A Marriage Sermon. A Sermon Called A Wife Mistaken.* London, 1641.

Graunt, John. *Natural and Political Observations Mentioned in a following Index, and made upon the Bills of Mortality.* London: Tho: Roycroft, 1662.

Graunt, John. *Reflections on the Weekly Bills of Mortality For the Cities of London and Westminster.* London: Samuel Speed, 1665.

Guild, William. *The Throne of David, or An Exposition of the Second of Samuell.* Oxford: W. Hall, 1659.

Hale, Matthew. *The Primitive Origination of Mankind, Considered and Examined According to the Light of Nature.* London: William Godbid, 1677

Halifax, George. *A Seasonable Address To both Houses of Parliament Concerning the Succession, The Fears of Popery, and Arbitrary Government.* London, 1681.

Hall, John. *Of government and obedience as they stand directed and determined by Scripture and reason.* London: T. Newcomb, 1654.

Harpur, John. *The ievvell of arithmetick.* London: Felix Kyngston, 1617.

Hawkins, John. *Clavis Commercii: or, The Key of Commerce.* London: Sarah Passinger, 1689.

Hearne, Thomas. *Ductor Historicus: Or, A Short System of Universal History and an Introduction to the Study of that Science Containing a Chronology of the most Celebrated Persons and Actions from the Creation to this Time.* London: Tim Childe, 1698.

de Henley, Walter. *The Booke of Thrift, Containing a Perfite Order, and Right Methode to Profite Lands, And Other Things Belonging To Husbandry; Newly Englished, and Set Out By I. B. Gentleman of Caen in France.* London: Iohn Wolfe, 1589.

Herbert, George. *A Priest to the Temple, Or, The Countrey Parson His Character, and Rule of Holy Life.* London: T. Moxey, 1652.

Hobbes, Thomas. *Leviathan, or, The matter, forme, and power of a common wealth, ecclesiasticall and civil.* London: Andrew Crooke, 1651.

Hodder, James. *Hodder's Arithmetick, or, That Necessary Art Made Most Easie.* London, 1663.

Hodder, James. *Hodder's Arithmetick: or, That Necessary Art Made Most Easie.* London: J. Dover, 1664.

Hodder, James. *Hodder's Arithmetick: or, That Necessary Art Made Most Easie.* London: J. Darby, 1667.

Hodder, James. *Hodder's Arithmetick, or, That Necessary Art Made Most Easie.* London: J. R., 1671.

Hodder, James. *Hodder's Arithmetick: or, That Necessary Art Made Most Easie.* London: T. J., 1672.

Hodder, James. *Hodder's Arithmetick, Or, That Necessary Art Made Most Easie.* Edited by Henry Mose. London: T. Hodgkin, 1678.

Hodder, James. *Hodder's Arithmetick: Or, That Necessary Art Made Most Easie.* Edited by Henry Mose. London: Ric. Chiswell, 1681.

Hodder, James. *Hodder's Arithmetick: or, That Necessary Art Made Most Easie.* Edited by Henry Mose. London: Ric. Chiswell and Tho. Sawbridge, 1683.

Hodder, James. *Hodder's Arithmetick: or, That Necessary Art Made Most Easie.* Edited by Henry Mose. London: Ric. Chiswell, 1685.

Hodder, James. *Hodder's Arithmetick: or, That Necessary Art Made Most Easie.* Edited by Henry Mose. London: Ric. Chiswel, 1690.

Hodder, James. *Hodder's Arithmetick, or, That Necessary Art Made Most Easie.* Edited by Henry Mose. London: Ric. Chiswell, 1693.

Hodder, James. *Hodder's Arithmetick. Or, That Necessary Art Made Most Easie.* Edited by Henry Mose. London: Ric. Chiswel, 1694.

Hodder, James. *Hodder's Arithmetick: or, That Necessary Art Made Most Easie.* Edited by Henry Mose. London: Ric. Chiswell, 1697.

Hodder, James. *Hodder's Decimal Arithmetick.* London: J. C., 1668.

Hodder, James. *Hodder's Decimal Arithmetick.* London: J. C., 1671.

Hodgson, James. *An Introduction to Chronology.* London: J. Hinton, 1747.

Hoole, Charles. *The Petty-Schoole, Shewing A Way to Teach Little Children to Read English with Delight and Profit, (especially) According to the New Primer.* London: J. T. for Andrew Crook, 1659.

Holme, Randle. *The Academy of Armory, or A Storehouse of Armory and Blazon.* Chester: Randle Holme, 1688.

Hopton, Arthur. *Hopton. 1606. An almanack and prognostication for this the second yeare after leape yeare.* London: Company of Stationers, 1606.

Hopton, Arthur. *Hopton 1611 An Almanack and Prognostication.* London: Company of Stationers, 1611.

Hopton, Arthur. *Hopton 1612. An Almanack and Prognostication.* London: Company of Stationers, 1612.

Horman, William. *Vulgaria uiri doctissimi.* London: Richard Pynson, 1519.

Houghton, John. *A Proposal for the Improvement of Husbandry and Trade.* London, 1691.

Howell, James. *Londinopolis; An Historicall Discourse or Perlustration Of the City of London.* London: J. Streater, 1657.

Howell, James. *Proedria Vasilike, A Discourse Concerning the Precedency of Kings.* London: James Cottrel, 1664.

Howell, William. *An Institution of General History from the beginning of the World to the Monarchy of Constantine the Great.* London: Henry Herringman, 1661.

Howes, Edward. *Short arithmetick: or, The old and tedious way of numbering, reduced to a new and briefe method.* London: R. Leybourne, 1650.

Huloet, Richard. *Hiloets Dictionarie.* Edited by John Higgins. London: Thomas Marsh, 1572.

Huarto de Mendoza, Diego. *The pleasant adventures of the witty Spaniard, Lazarillo de Tormes.* London: J. Leake, 1688.

Hunt, Nicolas. *The merchants ievvell: or, A new inuention arithmeticall.* London: Augustine Mathevves, 1632.

Hunt, Nicolas. *The hand-maid to arithmetick refined: shewing the variety and facility of working all rules in whole numbers and fractions.* London: Iohn Beale, 1633.

Hylles, Thomas. *The Arte of Vulgar Arithmeticke, both in Integers and Fractions.* London: Gabriel Simson, 1600.

Jackson, Arthur. *A Help For The Understanding of the Holy Scripture . . . Containing certain short notes of exposition upon the five books of Moses.* London: Roger Daniel, 1643.

Jager, Robert. *Artificial arithmetick in decimals*. London: Robert and William Leybourn, 1651.

Johnson, John. *Iohnsons Arithmatick in 2. bookes*. London: Augustine Mathewes, 1623.

Johnson, John. *Iohnsons Arithmatick in 2. bookes*. London: Augustine Mathewes and Robert Milborne, 1633.

Johnson, John. *Johnsons arithmetick in two books*. London: J. Flesher, 1671.

Johnson, John. *Johnson's arithmetick: in two books*. London: Andrew Clark, 1677.

Keble, Joseph. *The Statutes at Large in Paragraphs and Sections or Numbers, from Magna Charta Until This Time*. London: John Bill, Thomas Newcomb, Henry Hills, Richard Atkins, and Edward Atkins, 1681.

King, Gregory. *Natural and Political Observations on the State of England*. London, 1802.

L'Estrange, Roger. *A Caveat to the Cavaliers: or, An Antidote Against Mistaken Cordials*. London: Henry Brome, 1661.

de Laune, Thomas. *The Present State of London: or, Memorials Comprehending A Full and Succinct Account of the Ancient and Modern State Thereof*. London: George Larkin for Enoch Prosser and John How, 1681.

Latimer, Hugh. *The seconde Sermon of Maister Hughe Latimer, whych he preached before the Kynges maiestie, w[ith]in his graces Palayce at Westminster*. London: Ihon Day, 1549.

Leybourn, William. *Arithmetick, vulgar, decimal, and instrumental*. London: R. and W. Leybourn, 1657.

Leybourn, William. *Arithmetick: vulgar, decimal, instrumental*. London: J. Streater, 1668.

Leybourn, William. *Arithmetick: vulgar, decimal, instrumental, algebraical*. London: Thomas James, 1678.

Leybourn, William. *Cursus mathematicus. Mathematical sciences, in nine books. Comprehending arithmetick, vulgar, decimal, instrumental, algebraical . . .* London: Thomas Basset, Benjamin Tooke, Thomas Sawbridge, Awnsham and John Churchill, 1690.

Lilly, William. *Merlini Anglici Ephemeris: Or, Astrological Judgments for the year 1677*. London: J. Macock, 1677.

Livy, *The Romane Historie Written by T. Livius of Padua . . . Translated out of Latine into English, by Philemon Holland*. London: Adam Islip, 1600.

Lupton, Donald. *A Warre-like Treatise of the Pike, or, Some Experimentall Resolves, for lessening the number, and disabling the use of the Pike in Warre*. London: Richard Hodgkinsonne, 1642.

Lyte, Henry. *The art of tens, or Decimall arithmeticke*. London: Edward Griffin, 1619.

M., A. *The Description Of a Plain Instrument*. London: J. Coniers, 1668.

Maidwell, Lewis. *A Scheme for a Public Academy*. London, 1700.

Maidwell, Lewis. *An Essay Upon the Necessity and Excellency of Education*. London: S. B. and J. B., 1705.

Makin, Bathsua. *An Essay to Revive the Antient Education of Gentlewomen, in Religion, Manners, Arts & Tongues*. London: J. D., 1673.

Malynes, Gerard. *Consuertudo, vel, Lex Mercatoria*. London: T. Basset, 1686.

Malynes, Gerard. *Englands View, in the Vnmasking of Two Paradoxes*. London: Richard Field, 1603.

Manby, Thomas. *A collection of the statutes made in the reigns of King Charles the I. and King Charles the II. with the abridgment of such as stand repealed or expired*. London: John Streater, James Flesher, and Henry Twyford, 1667.

Markham, Gervase. *The English Husbandman, drawn into two Bookes, and each Booke into two Parts*. London: William Sheares, 1635.

Martindale, Adam. *The Countrey-Survey-Book: or Land-Meters Vade Mecum.* London: A. G. and J. P., 1692.

Masterson, Thomas. *Thomas Masterson his first booke of arithmeticke.* London: Richard Field, 1592.

Masterson, Thomas. *Thomas Masterson his addition to his first booke of arithmetick.* London: Richard Field, 1594.

Masterson, Thomas. *Thomas Masterson his third booke of arithmeticke.* London: Richard Field, 1595.

Masterson, Thomas. *Masterson's arithmetick.* Edited by Humfrey Waynman. London: George Miller, 1634.

Mayne, John. *Arithmetick: Vulgar, Decimal, & Algebraical.* London: J. A., 1675.

Mead, Matthew. *Solomon's Prescription For the Removal of the Pestilence: or, The Discovery of the Plague of our Hearts, in order to the Healing of that in our Flesh.* London, 1665.

Moore, Jonas. *Moores arithmetick: discovering the secrets of that art, in number and species. In two books.* London: Thomas Harper, 1650.

Moore, Jonas. *Moores arithmetick. In two bookes.* London: Thomas Harper, 1650.

Moore, Jonas. *Moor's arithmetick. In tvvo books.* London: J. G., 1660.

Moore, Jonas. *A Mathematical Compendium; or, Useful practices in arithmetick, geometry, and astronomy, geography and navigation, embatteling, and quartering of armies, fortification and gunnery, gauging and dyalling.* Edited by Nicholas Stephenson. London: Robert Harford, 1681.

Moore, Jonas. *Moore's Arithmetick: in four books.* London: Ralph Holt, 1688.

Moore, Jonas. *A mathematical compendium; or, useful practices in arithmetick, geometry, and astronomy . . . .* Edited by R. H. London: J. Philip, H. Rhodes, and J. Taylor, 1705.

Morland, Samuel. *The description and use of two arithmetick instruments. Together with a short treatise, explaining and demonstrating the ordinary operations of arithmetick.* London: Moses Pitt, 1673.

Moxon, Joseph. *A Tutor to Astronomie and Geographie.* London, 1659.

Mulcaster, Richard. *The First Part of the Elementarie Which Entreateth Chefelie of the right writing of our English tung.* London: Thomas Vautroullier, 1582.

Mulcaster, Richard. *Positions Wherin Those Primitiue Circumstances Be Examined, Which Are Necessarie for the Training Vp of Children.* London: Thomas Vautroullier, 1581.

Nedham, Marchamont. *Medela Medicinae. A Plea For the Free Profession, and a Renovation of the Art of Physick, Out of the Noblest and most Authentick Writers.* London: Richard Lownds, 1665.

Newbold, A. *Londons Improvement, and the Builders Security, &c.* London: Thomas Milbourn, 1680.

North, Roger. *The Gentleman Accomptant: OR, AN ESSAY To unfold the Mystery of Accompts.* London: E. Curll, 1714.

Oldcastle, Hugh. *A briefe instruction and maner hovv to keepe bookes of accompts after the order of debitor and creditor.* Edited by John Mellis. London: Iohn Windet, 1588.

Orpen, Richard. *An Exact Relation of the Persecutions, Robberies, and Losses, sustained by the Protestants of Killmare, in Ireland.* London: Thomas Bennet, 1689.

Palsgrave, John. *Lesclarcissement de la Langue Francoyse.* London: Richard Pynson and Iohan Haukyns, 1530.

Parker, Martin. *Lord Have Mercy Upon Us.* London: Thomas Lambert, 1636.

Patrick, Simon. *A Paraphrase Upon the Books of Ecclesiastes and the Song of Solomon with arguments to each chapter and annotations thereupon.* London: W. H., 1700.

Pecke, Thomas. *Parnassi Puerperium.* London: James Cottrel, 1659.

Perkins, Francis. *Perkins. A New Almanack and Prognostication for the Year of Our Lord God 1671*. London: E. L. and Robert White, 1671.

Perkins, William. *A Commentarie, or Exposition, vpon the fiue first Chapters of the Epistle to the Galatians*. Cambridge: John Legat, 1604.

Perkins, William. *Fovre Great Lyers, Striuing who shall win the siluer Whetstone*. London: Robert Walde-graue, 1585.

Petri, Nicolaus. *The Pathway to Knowledge. . . . How to Cast Accompt with Counters, and with Pen, Both in Whole, and Broken Numbers*. Translated by William Phillip. London: Abel Jeffes, 1596.

Petty, William. "An Extract of two Essays in Political Arithmetick concerning the comparative Magnitudes &c. of London and Paris by Sir William Petty Knight, R.S.S." *Philosophical Transactions of the Royal Society* 16 (1686).

Petty, William. *England's Guide to Industry: or, Improvement of Trade for the good of all People in general*. London: R. Holt, 1683.

Petty, William. *Further Observation Upon the Dublin-Bills: Or, Accompts of the Houses, Hearths, Baptisms, And Burials in that City*. London: Mark Pardoe, 1686.

Petty, William. *Political Arithmetick, or A Discourse Concerning The Extent and Value of Lands, People, Buildings. . . .* London: Robert Clavel, 1690.

Philipps, Fabian. *The Pretended Perspective-Glass*. London: s.n., 1669.

Pond, Edward. *Pond's Almanack for the Yeare of Our Lord God 1641*. Cambridge: Rog. Daniel, 1641.

Pratt, William. *The Arithmeticall Jevvell*. London: Iohn Beale, 1617.

Primaudaye, Pierre de La. *The French Academie Fully discoursed and finished in foure bookes*. London: Thomas Adams, 1618.

Prynne, William. *A Briefe Polemicall Dissertation, Concerning the True Time of the Ichoation and Determination of the Lordsday-Sabbath*. London: T. Mabb, 1654.

Purchas, Samuel. *Purchas His Pilgrimes. In Five Bookes. The third, Voyages and Discoueries of the North parts of the World, by Land and Sea, in Asia, Evrope, the Polare Regions, and in the North-west of America*. London: William Stansby for Henri Fetherstone, 1625.

Randolph, Thomas. *Poems with the Muses looking-glasse*. Oxford: Leonard Lichfield, 1638.

Rawlyns, Richard. *Practical Arithmetick in Whole Numbers. Fractions. Decimals*. London: Humphrey Mosely and Richard Tomlins, 1656.

Raworth, Francis. *Blessedness, or, God and the World Weighed in the Balances of the Sanctuary and the World found too light. Preached in a Sermon at Pauls*. London: T. Maxey, 1656.

Recorde, Robert. *The Ground of Artes Teachyng the Worke and Practise of Arithmetike*. London: Reynold Wolfe, 1543.

Recorde, Robert. *The Ground of Artes Teachyng the Worke and Practise of Arithmetike*. London: S.I. for R. Wolff, 1549.

Recorde, Robert. *The Ground of Artes*. London: Reynold Wolff, 1552.

Recorde, Robert. *The Grounde of Artes*. Edited by John Dee. London: Reginalde VVolfe. 1561.

Recorde, Robert. *The Grounde of Artes*. Edited by John Dee. London: Reginalde Wolfe, 1566.

Recorde, Robert. *The Ground of Arts*. Edited by John Dee. London: R. Wolfe, 1573.

Recorde, Robert. *The Grounde of Artes*. Edited by John Dee. London: Henry Binneman and John Harison, 1575.

Recorde, Robert. *The Grounde of Artes*. Edited by John Dee. London: H. Bynneman, 1579.

Recorde, Robert. *The Grounde of Artes*. Edited by John Dee and John Mellis. London: I. Harison and H. Bynneman, 1582.

Recorde, Robert. *The Ground of Artes*. Edited by John Dee and John Mellis. London: Henrie Midleton, 1586.

Recorde, Robert. *The Ground of Artes*. Edited by John Dee and John Mellis. London: T. Dawson, 1594.

Recorde, Robert. *The Ground of Artes*. Edited by John Dee and John Mellis. London: Richard Field, 1596.

Recorde, Robert. *The Ground of Arts*. Edited by John Dee and John Mellis. London: I. Harrison, 1600.

Recorde, Robert. *The Grounde of Artes*. Edited by John Dee and John Mellis. London: N. Okes, 1607.

Recorde, Robert. *The Ground of Artes*. Edited by John Dee, John Mellis, and John Wade. London: William Hall, 1610.

Recorde, Robert. *Records Arithmeticke: Contayning the Ground of Arts*. Edited by John Dee, John Mellis, and Robert Norton. London: Thomas Snodham, 1615.

Recorde, Robert. *The Ground of Arts*. Edited by John Dee, John Mellis, Robert Norton, and Robert Hartwell. London: Iohn Beale, 1618.

Recorde, Robert. *The Grounde of Artes*. Edited by John Dee, John Mellis, and Robert Harwell. London: John Beale, 1623.

Recorde, Robert. *The Grounde of Artes*. Edited by John Dee, John Mellis, Robert Norton, and Robert Hartwell. London: Thomas Harper, 1631.

Recorde, Robert. *The Ground of Artes*. Edited by John Dee, John Mellis, and Robert Hartwell. London: Thomas Harper, 1632.

Recorde, Robert. *The Ground of Arts*. Edited by John Dee, John Mellis, and Robert Hartwell. London: Thomas Harper, 1636.

Recorde, Robert. *The Ground of Arts*. Edited by John Dee, John Mellis, and Robert Hartwell. London: J. Raworth and J. Okes, 1640.

Recorde, Robert. *The Ground of Arts*. Edited by John Dee, Jon Mellis, and Robert Hartwell. London: Miles Flesher, 1646.

Recorde, Robert. *Records Arithmetick: or, The ground of arts*. Edited by John Dee, John Mellis, and Robert Hartwell. London: Miles Flesher, 1648.

Recorde, Robert. *Records Arithmetick: or, The ground of arts*. Edited by John Dee, John Mellis, and Robert Hartwell. London: J. Flesher, 1652.

Recorde, Robert. *Record's Arithmetick, or, the Ground of Art*. Edited by John Dee, John Mellis, and Robert Hartwell. London: James Flesher, 1654.

Recorde, Robert. *Records Arithmetick: or, The ground of arts*. Edited by John Dee, John Mellis, and Robert Hartwell. London: James Flesher, 1658.

Recorde, Robert. *Records Arithmetick: or, The ground of arts*. Edited by John Dee, John Mellis, and Robert Hartwell. London: James Flesher, 1662.

Recorde, Robert. *Record's Arithmetick, or, The ground of arts*. Edited by John Dee, John Mellis, and Robert Hartwell. London: James Flesher, 1668.

Recorde, Robert. *Record's Arithmetick: or, The ground of arts*. Edited by John Dee, John Mellis, and Robert Hartwell. London: E. Flesher, 1673.

Recorde, Robert. *Arithmetick: or, The Ground of Arts: Teaching that Science, Both in Whole Numbers and Fractions*. Edited by Edward Hatton. London: John Heptinstall, 1699.

Recorde, Robert. *The Whetstone of Witte, whiche is the seconde parte of Arithmetike*. London: Ihon Kyngstone, 1557.

Rich, Barnabe. *A nevv description of Ireland vvherein is described the disposition of the Irish whereunto they are inclined.* London: William Jaggard, 1610.

Roberts, Lewis. *The Merchants Mappe of Commerce: Wherein The Vniversall Manner and Matter of Trade, is compendiously handled.* London: R.O for Ralh Mabb, 1638.

Sanderson, William. *A Compleat History of the Life and Raigne of King Charles From His Cradle to his Grave.* London: Humphrey Moseley, Richard Tomlins, George Sawbridge, 1658.

Savage, William. *Savage 1611. A new Almanacke and Prognostication, for the yeare of our Lord and Sauiour Jesus Christ, 1611.* London: Company of Stationers, 1611.

Scot, Thomas. *Philomythie, or Philomythologie wherein Outlandish Birds, Beasts, and Fishes, are taught to Speake true English plainely.* London: Francis Constable, 1622.

Seaman, Henry. *Kalendarium Nauticum: The Sea-man's Almanack.* London: T. N., 1677.

Seller, John. *Jamaica Almanack.* London, 1684.

Sergeant, John. *Faith vindicated from possibility of falshood, or, The immovable firmness and certainty of the motives to Christian faith.* Lovain, 1667.

Sergeant, John. *A vindication of the doctrine contained in Pope Benedict XII, his bull and in the General Council of Florence, under Eugenius the iiii concerning the state of departed souls.* Paris, 1659.

Shakespeare, William. *The First Part of the Contention Betwixt the Two Famous Houses of Yorke and Lancaster with the Death of the Good Duke Humphrey: and the Banishment and Death of the Duke of Suffolke, and the Tragical End of the Prowd Cardinall of Winchester, with the Notable Rebellion of Iacke Cade: and the Duke of Yorkes First Clayme to the Crowne.* London: Valentine Simmes for Thomas Millington, 1600.

Shepard, Thomas. *Theses Sabbaticae, or, The Doctrine of the Sabbath.* London, 1650.

Sheppard, William. *The Touch-Stone of Common Assurances.* London: M. F. for W. Lee, M. Walbancke, D. Pakeman, and G. Bedell, 1648.

Shiers, William. *A Familiar Discourse or Dialogue Concerning the Mine-Adventure.* London: 1700.

Sincera, Rege. *Observations both Historical and Moral Upon the Burning of London.* London: Thomas Ratcliffe, 1667.

Smith, John. *The Generall Historie of Virginia, New-England, and the Summer Isles with the names of the Adventurers, Planters, and Governours from their first beginning. Ano: 1584 to this present 1624.* London: I. D. and I. H. for Michael Sparkes, 1624.

Smythe, John. *Certen Instructions, Obseruations and Orders Militarie, Requisit for all Chieftaines, Captaines, and Higher and Lower Men of Charge, and Officers, to Vnderstand, Knowe and Obserue.* London: Richard Iohnes, 1594.

Spence, Craig. *Accidents and Violent Death in Early Modern London: 1650–1750.* Woodbridge, Suffolk: Boydell Press, 2016.

Stevin, Simon. *Disme: the Art of Tenths, or, Decimall Arithmetike.* Translated by Robert Norton. London: S. Stafford, 1608.

Stow, John, and Edmund Howes. *Annales, or A Generall Chronicle of England.* London: Richardi Meighen, 1631.

Strode, Thomas. *An Arithmetical Treatise of the Combinations, Elections, Permutations, and Composition, of Quantities.* London: John Taylor, 1693.

Temple, William. *Observations upon the United Provinces of the Netherlands.* London: A Maxwell, 1673.

Tunstall, Cuthbert. *De Arte Supputandi Libri Quattuor.* London: Richard Pynson, 1522.

Tunstall, Cuthbert. *De Arte Svppvtandi Libri Qvatvor.* Paris: Robert Stephan, 1529.

Tunstall, Cuthbert. *De Arte Svpputandi Libri Quttuor.* Paris: Robert Stephan, 1538.

Vernon, John. *The Compleat Comptinghouse: OR, The young Lad taken from the Writing School, and fully instructed, by way of Dialogue, in all the Mysteries of a Merchant*. London: J. D. for Benj. Billingsley, 1678.

Wase, Christopher. *Considerations Concerning Free-Schools as Settled in England*. London: Simon Millers, 1678.

Webster, William. *The Principles of Arithmetick*. London: Miles Flesher, 1634.

Wentworth, Thomas. *The Office and Dutie of Executors*. London: T. C. for Andrew Crooke, Laurence Chapman, William Cooke, and Richard Best, 1641.

West, William. *The First Part of Simboleography*. London: Companie of Stationers, 1610.

Wharton, George. *Calendarium ecclesiasticum: or A New Almanack after the Old Fashion*. London: John Grismond, 1659.

Wharton, George. *No Merline, nor Mercury: but a New Almanack after the Old Fashion*. London, 1648.

White, John. *The Country-Man's Conductor in reading and writing true English . . . and some arithmetical rules to be learnt by children, before or as soon as they are put to Writing*. Exeter: Samuel Farley, 1701.

Whitlock, Richard. *Zootomia, or Observations on the Present Manners of the English*. London: Tho. Roycroft, 1654.

Wilkins, John. *Of the principles and duties of natural religion*. London: A. Maxwell, 1675.

Wilkinson, Thomas. *Wilkinson, 1657. A Kelander and Prognostication*. London: G. Dawson, 1657.

Willet, Andrew. *An Harmonie. Vpon the Second Booke of Samvuel*. Cambridge: Cantrell Legge, 1614.

Willet, Andrew. *Hexapia in Genesin & Exodum: That is, A sixfold commentary upon the two first Bookes of Moses, being Genesis and Exodvs*. London: John Haviland, 1633.

Williams, William. *Votes of the House of Commons, Perused and Signed to be Printed According to the Order of the House of Commons*. London: 1680.

Willsford, Thomas. *Willsfords Arithmetick, Naturall, and Artificiall: or, Decimalls*. London: John Grismond, 1656.

Wilson, Thomas. *A Discourse Vppon Vsurye*. London: Rychardi Tottelli, 1572.

Wing, Vincent. *Wing's Ephemeris for Thirty Years*. London: J. C., 1669.

Wingate, Edmund. *Arithmetique Made Easie*. London: Miles Flesher, 1630.

Wingate, Edmund. *Arithmetique Made Easie*. Edited by John Kersey. London: J. Flesher, 1650.

Wingate, Edmund. *Arithmetique Made Easie, the Second Book*. Edited by John Kersey. London: R. & W. Leybourn, 1652.

Wingate, Edmund. *Mr. Wingate's Arithmetick*. Edited by John Kersey. London: Philemon Stephens, 1658.

Wingate, Edmund. *Mr. Wingate's Arithmetick*. Edited by John Kersey. London: Thomas Roycroft, 1668.

Wingate, Edmund. *Mr. Wingate's Arithmetick*. Edited by John Kersey. London: Thomas Roycroft, 1673.

Wingate, Edmund. *Mr. Wingate's Arithmetick*. Edited by John Kersey. London: Samuel Roycroft, 1678.

Wingate, Edmund. *Mr. Wingate's Arithmetick*. Edited by John Kersey. London: Edward Horton, 1683.

Wingate, Edmund. *Mr. Wingate's Arithmetick*. Edited by John Kersey. London: John Williams, 1689.

Wingate, Edmund. *Mr. Wingate's Arithmetick.* Edited by John Kersey. London: John Williams, 1694.

Wingate, Edmund. *Mr. Wingate's Arithmetick.* Edited by John Kersey. London: J. Philips, 1699.

Wingate, Edmund. *The Clarks Tutor for Arithmetick and Writing.* Edited by Edward Cocker. London: S. Griffin, 1670.

Wingate, Edmund. *The Clarks Tutor for Arithmetick and VVriting.* Edited by Edward Cocker. London: S. Griffin, 1671.

Wingate, Edmund. *Wingate's Remains: or, The Clerks Tutor to Arithmetick and Writing.* Edited by Edward Cocker. London: Henry Twyford, 1676.

Wood, Robert. *A New Almonac for Ever, or A Rectified Account of Time.* London, 1680.

Wood, Robert. *A Specimen of a New Almonac for Ever: or A Rectified Account of Time.* London, 1680.

Woodhouse, William. *Woodhouse. 1607. An Almanacke, and Prognostication.* London: Company of Stationers, 1607.

Worsop, Edward. *A Discouerie of sundrie errours and faults daily committed by Lande-Meaters, ignorant of Arithmetike and Geometrie.* London: Henrie Middleton for Gregorie Seton, 1582.

Wurstisen, Christian. *The elements of arithmeticke most methodically deliuered.* Translated by Thomas Hood. London: Richard Field, 1596.

Modern Print Editions of Sources

Adams, Simon, ed. *Household Accounts and Disbursement Books of Robert Dudley, Earl of Leicester, 1558–1561, 1584–1586.* London: Cambridge University Press, 1995.

Akkerman, Nadine, ed. *The Correspondence of Elizabeth Stuart, Queen of Bohemia,* vol. 2, *1632-1642.* Oxford: Oxford University Press, 2011.

Ascham, Roger. *Toxophilus.* Tempe: Arizona Center for Medieval and Renaissance Studies, 2002.

Bailey, F. A., ed. *The Churchwardens' Accounts of Prescot, Lancashire, 1523-1607.* Preston: Record Society of Lancashire and Cheshire, 1953.

Bankes, Joyce, and Eric Kerridge, eds. *The Early Records of the Bankes Family at Winstanley.* Remains, Historical and Literary, connected with the Palatine Counties of Lancaster and Chester, 3rd ser., 21. Manchester: Manchester University Press, 1973.

Bernoulli, Jacob. *The Art of Conjecturing together with Letter to a Friend on Sets in Court Tennis.* Translated by Edith Dudley Sylla. Baltimore: Johns Hopkins University Press, 2006.

Bonsey, Carol G., and J. G. Jenkins, eds. *Ship Money Papers and Richard Grenville's Note-Book.* Buckinghamshire Record Society Publications 13. Hertfordshire: Broadwater Press, 1965.

Boyd, William K., and Henry W. Meikle, eds. *Calendar of State Papers relating to Scotland and Mary, Queen of Scots, 1547-1603,* vol. 10, *1589-1593.* Edinburgh, Scotland: H.M. General Register House, 1936.

Brewer, J. S. ed. *Letters and Papers, Foreign and Domestic, of the Reign of Henry VIII,* vol. 3, part 1, *1519-21.* London: Longman, Green, Reader and Dyer, 1867.

Bruce, John, ed. *Correspondence of King James VI. of Scotland with Sir Robert Cecil and others in England, during the reign of Queen Elizabeth; with an appendix containing papers illustrative of transactions between King James and Robert Earl of Essex.* [Westminster]: Camden Society, 1861.

Burgess, Clive, ed. *The church records of St Andrew Hubbard, Eastcheap, c1450–c1570.* [London]: London Record Society, 1999.

Byrne, Muriel St. Clare, ed. *The Lisle Letters.* Chicago: University of Chicago Press, 1981.

Cressy, David, and Lori Anne Ferrell. *Religion and Society in Early Modern England: A Sourcebook*, 2nd ed. New York: Routledge, 2005.

Croston, James, ed. *The Register Book of Christenings, Weddings, and Burials within the Parish of Prestbury, in the County of Chester, 1560–1636.* Record Society of Lancashire and Cheshire 5. Manchester: A. Ireland, 1881.

Defoe, Daniel. *A Journal of the Plague Year.* London: Falcon Press, 1950.

Doree, Stephen G., ed. *The Parish Register and Tithing Book of Thomas Hassall of Amwell.* [Ware, Hertfordshire]: Hertfordshire Record Society, 1989.

Earwaker, J. P. ed. *Lancashire and Cheshire Wills and Inventories at Chester.* Remains, Historical and Literary, connected with the Palatine Counties of Lancaster and Chester, n.s., 3. Manchester: Charles E. Simms, 1884.

Earwaker, J. P., ed. *Lancashire and Cheshire Wills and Inventories, 1572 to 1696, Now Preserved at Chester.* Remains, Historical and Literary, connected with the Palatine Counties of Lancaster and Chester, n.s., 28. Manchester: Charles E. Simms, 1893.

Firth, C. H. *Acts and Ordinances of the Interregnum, 1642–1660.* London: H. M. Stationery Office, 1911.

Fussell, G. E. *Robert Loder's Farm Accounts, 1610–1620.* Camden Third Series, 53. London: Butler and Tanner, 1936.

Gairdner, J., ed. *Letters and Papers, Foreign and Domestic, of the Reign of Henry VIII*, vol. 13, part 2, *1538.* London: Her Majesty's Stationery Office, 1893.

Gardiner, Dorothy, ed. *The Oxinden Letters, 1607–1642; being the correspondence of Henry Oxinden of Barham and his circle.* London: Constable, 1933.

Gardiner, Dorothy, ed. *The Oxinden and Peyton Letters, 1642–1670. Being the correspondence of Henry Oxinden of Barham, Sir Thomas Peyton of Knowlton.* London: Sheldon Press, 1937.

Greaves, R. W. ed. *The First Ledger Book of High Wycombe.* Buckinghamshire Record Society Publications 11. Hertfordshire: Broadwater Press, 1947.

Griffiths, Ralph A., ed. *The Household Book (1510–1551) of Sir Edward Don: An Anglo-Welsh Knight and his Circle.* Buckinghamshire Record Society Publications 33. Amersham, Buckinghamshire: Buckinghamshire Record Society, 2004.

Hendricks, Frederick. *Contributions to the History of Insurance, and of the Theory of Life Contingencies, with a Restoration of the Grand Pensionary De Wit's Treatise on Life Annuities.* London: C. and E. Layton, 1851.

Hickman, David, ed. *Lincoln Wills 1532–1534.* Publications of the Lincoln Record Society, 89. Suffolk: Boydell Press, 2001.

King, Ann J., ed. *Muster Books for North and East Hertfordshire, 1580–1605*, Hertfordshire Record Publications 12. Hitchin: Hertfordshire Record Society, 1996.

Kirby, Joan, ed. *The Plumpton Letters and Papers.* Camden Fifth Series 8. Cambridge: Cambridge University Press, 1996.

Legg, L. G. Wickham, ed. *A Relation of A Short Survey of 26 Counties Observed in a seven weeks Journey begun on August 11, 1634 By a Captain, a Lieutenant, and an Ancient All three of the Military Company in Norwich.* London: F. E. Robinson, 1904.

Martindale, Adam. *The Life of Adam Martindale Written By Himself.* Edited by Richard Parkinson. Otley, West Yorkshire: Chetham Society, 2001.

Mellows, William., ed. *Peterborough local administration: parochial government before the reformation. Churchwardens' accounts, 1467-1573, with supplementary documents, 1107-1488.* Publications of the Northamptonshire Record Society 9. Kettering: Northamptonshire Record Society, 1939.

Mellows, William., ed. *Peterborough local administration: parochial government from the Reformation to the Revolution, 1541-1689. Minutes and accounts of the feoffees and governors of the city lands with supplementary documents.* Publications of the Northamptonshire Record Society 10. Kettering: Northamptonshire Record Society, 1937.

Northbrooke, John. *A Treatise Against Dicing, Dancing, Plays, and Interludes, From the Earliest Edition, About A.D. 1577.* London: Shakespeare Society, 1843.

Palmer, Anthony, ed. *Tudor Churchwardens' Accounts.* Hertfordshire Record Society, 1 [Ware, Hertfordshire]: Hertfordshire Record Society, 1985.

Pepys, Samuel. *The Diary of Samuel Pepys.* Edited by Robert Latham and William Matthews. Berkeley: University of California Press, 1970-1983.

Pound, John F., ed. *The Norwich Census of the Poor, 1570,* Norfolk Record Society 40. Norfolk: Norfolk Record Society, 1971.

Power, Michael., ed. *Liverpool Town Books, 1649-1671.* Record Society of Lancashire and Cheshire 136. Liverpool: Record Society of Lancashire and Cheshire, 1999.

Ramsay, Esther M. E., ed. *The churchwardens' accounts of Walton-on-the-Hill, Lancashire 1627-1667.* [Liverpool]: The Record Society of Lancashire and Cheshire, 2005.

Ramsay, Esther M. E., ed. *John Isham, mercer and merchant adventurer: two account books of a London merchant in the reign of Elizabeth I.* Gateshead, Co. Durham: Northamptonshire Record Society, 1962.

Recorde, Robert. *The Grounde of Artes Teachyng the Worke and Practise of Arithmeticke.* London: Reynold Wolff, 1542; Amsterdam: Theatrum Orbis Terrarum, 1969.

Roberts, R.A., ed. *Calendar of the Manuscripts of the Most Hon. the Marquis of Salisbury, Preserved at Hatfield House, Hertfordshire,* vol. 10, *1600.* London: His Majesty's Stationery Office, 1904.

Rylands, John Paul, ed. *Cheshire and Lancashire Funeral Certificates, 1600 to 1678.* Record Society of Lancashire and Cheshire, 6. London: Wyman and Sons, 1882.

Savage, Richard, and Edgar L. Fripp, eds. *Minutes and accounts of the corporation of Stratford-upon-Avon and other records, 1553-1620.* Oxford: Dugdale Society, 1921.

Seaman, Paul, ed. *Norfolk Hearth Tax Exemption Certificates 1670-1674: Norwich, Great Yarmouth, King's Lynn and Thetford.* Norfolk Record Society 65. London: British Record Society, 2001.

Stocks, George Alfred, ed. *The Records of Blackburn Grammar School.* Remains, Historical and Literary, connected with the Palatine Counties of Lancashire and Chester, n.s., 66. Manchester: Charles E. Simms, 1909.

Tittler, Robert, ed. *Accounts of the Roberts Family of Boarzell, Sussex: c1568-1582.* Sussex Records Series, 71. Lewes, England: Sussex Record Society, 1977.

Weaver, F. W., and G. N. Clark, eds. *Churchwardens' Accounts of Marston, Spelsbury, Pyrton.* Oxfordshire Records Series, 6. Oxford: Oxford University Press, 1925.

Webb, C. C., ed. *The Churchwardens' Accounts of St Michael, Spurriergate, York, 1518-1548.* [York]: University of York, Borthwick Institute of Historical Research, 1997.

Whiteman, Anne, ed. *The Compton Census of 1676*, British Academy Records of Social and Economic History, n.s., 10. London: Oxford University Press, 1986.
Willis, Arthur J., and A. L. Merson, eds. *A Calendar of Southampton Apprenticeship Registers, 1609–1740*. Southampton: Southampton University Press, 1968.

Secondary Sources

Abboud, Sami, Shachar Maidenbaum, Stanislas Dehaene, and Amir Amedi. "A Number-Form Area in the Blind." *Nature Communications* 6, no. 1 (January 23, 2015): 1–9.
Agrillo, Christian. "Numerical and Arithmetic Abilities in Non-primate Species." In *The Oxford Handbook of Numerical Cognition*. Edited by Roi Cohen Kadosh and Ann Dowker, 214–36. Oxford: Oxford University Press, 2015.
Agrillo, Christian, and Angelo Bisazza. "Understanding the Origin of Number Sense: A Review of Fish Studies." *Philosophical Transactions of the Royal Society B: Biological Sciences* 373, no. 1740 (February 19, 2018): 20160511.
A'Hearn, Brian, Jörg Baten, and Dorothee Crayen. "Quantifying Quantitative Literacy: Age Heaping and the History of Human Capital." *Journal of Economic History* 69, no. 3 (September2009): 783–808.
Akin, Marjorie H., James C. Bard, and Kevin Akin. *Numismatic Archaeology of North America: A Field Guide to Historical Artifacts*. New York: Routledge, 2016.
Alexander, Amir. "The Skeleton in the Closet: Should Historians of Science Care about the History of Mathematics?" *Isis* 102 (2011): 475–80.
Alexander, Michael Van Cleave. *The Growth of English Education, 1348–1648: A Social and Cultural History*. University Park: Pennsylvania State University Press, 1990.
Al-Khalili, Jim. *The House of Wisdom: How Arabic Science Saved Ancient Knowlege and Gave Us the Renaissance*. New York: Penguin Press, 2011.
Alföldi-Rosenbaum, Elisabeth. "The Finger Calculus in Antiquity and in the Middle Ages: Studies on Roman Game Counters I." *Frühmittelalterliche Studien* 5, no 1 (1971): 1–9.
Altegoer, Diana B. *Reckoning Words: Baconian Science and the Construction of Truth in English Renaissance Culture*. Madison: Fairleigh Dickinson University Press, 2000.
Amalric, Marie, and Stanislas Dehaene. "A Distinct Cortical Network for Mathematical Knowledge in the Human Brain." *NeuroImage* 189 (April 1, 2019): 19–31.
Amalric, Marie, and Stanislas Dehaene. "Cortical Circuits for Mathematical Knowledge: Evidence for a Major Subdivision within the Brain's Semantic Networks." *Philosophical Transactions of the Royal Society B: Biological Sciences* 373, no. 1740 (February 19, 2018): 20160515.
Amalric, Marie, and Stanislas Dehaene. "Origins of the Brain Networks for Advanced Mathematics in Expert Mathematicians." *Proceedings of the National Academy of Sciences* 113, no. 18 (May 3, 2016): 4909–17.
Amalric, Marie, Isabelle Denghien, and Stanislas Dehaene. "On the Role of Visual Experience in Mathematical Development: Evidence from Blind Mathematicians." *Developmental Cognitive Neuroscience* 30 (April 1, 2018): 314–23.
Andres, Michael and Mauro Pesenti. "Finger-Based Representation of Mental Arithmetic." In *The Oxford Handbook of Numerical Cognition*. Edited by Roi Cohen Kadosh and Ann Dowker, 67–88. Oxford: Oxford University Press, 2015.

Angus, John. "Old and New Bills of Mortality; Movement of the Population; Deaths and Fatal Diseases in London during the Last Fourteen Years." *Journal of the Statistical Society of London* 17, no. 2 (1854): 117–42.

Archer, Ian W. "The Burden of Taxation on Sixteenth-Century London." *Historical Journal* 44, no. 3 (September 2001): 599–627.

Ascher, Marcia. *Ethnomathematics: A Multicultural View of Mathematical Ideas*. Pacific Grove, CA: Brooks/Cole, 1991.

Ash, Eric H. *Power, Knowledge and Expertise in Elizabethan England*. Baltimore: Johns Hopkins University Press, 2004.

Ashton, John. *The History of Gambling in England*. Chicago: Herbert S. Stone, 1899.

Baird, Caroline. *Games and Gaming in Early Modern Drama: Stakes and Hazards*. Cham, Switzerland: Palgrave Macmillan, 2020.

Barker, Peter, and Roger Ariew, ed. *Revolution and Continuity: Essays in the History and Philosophy of Early Modern Science*. Washington, DC: Catholic University of America Press, 1991.

Barnard, Francis Pierrepont. *The Casting-Counter and the Counting-Board: A Chapter in the History of Numismatics and Early Arithmetic*. Oxford: Clarendon Press, 1916.

Barnard, Francis Pierrepont. "Portuguese Jettons." *Numismatic Chronicle and Journal of the Royal Numismatic Society* 5th ser, 3 (1923): 75–114.

Barnard, John, and Maureen Bell. Appendix 1: Statistical Tables. In *The Cambridge History of the Book in Britain*, vol. 4, *1557–1695*. Edited by John Barnard and D. F. McKenzie, 779–85. Cambridge: Cambridge University Press, 2002.

Barrington, Daines. "An Historical Disquisition on the Game of Chess; addressed to Count de Bruhl, F.A.S." *Archaeologia* 9. London: J. Nichols, 1789.

Barry, Jonathan. "Popular Culture in Seventeenth-Century Bristol." In *Popular Culture in Seventeenth-Century England*. Edited by Barry Reay, 59–90. London: Croom Helm, 1985.

Baxter, W. T. "Early Accounting: The Tally and the Checkerboard." *Accounting Historians Journal* 16, no. 2 (December 1989): 43–83.

Beal, Peter. *In Praise of Scribes: Manuscripts and Their Makers in Seventeenth-Century England*. Oxford: Clarendon Press, 1998.

Belenkiy, Ari, and Eduardo Vila Echagüe. "History of One Defeat: Reform of the Julian Calendar as Envisaged by Isaac Newton." *Notes and Records of the Royal Society of London* 59, no. 3 (September 22, 2005): 223–54.

Bellhouse, D. R. "Probability in the Sixteenth and Seventeenth Centuries: An Analysis of Puritan Casuistry." *International Statistical Review* 56, no. 1 (April 1988): 63–74.

Ben-Amos, Ilana Krausman. *Adolescence and Youth in Early Modern England*. New Haven, CT: Yale University Press, 1994.

Bennett, Jim. "Early Modern Mathematical Instruments." *Isis* 102 (2011): 697–705.

Bennett, Jim. "Mathematics, Instruments and Navigation, 1600–1800." In *Mathematicians and the Historian's Craft: The Kenneth O. May Lectures*. Edited by Glen Van Brummelen and Michael Kinyon, 43–55. New York: Springer, 2005.

Bennett, Kate. "John Aubrey and the 'Lives of Our English Mathematical Writers.' " In *The Oxford Handbook of the History of Mathematics*. Edited by Eleanor Robson and Jacqueline Stedall, 329–52. Oxford: Oxford University Press, 2009.

Benson-Amram, Sarah, Geoff Gilfillan, and Karen McComb. "Numerical Assessment in the Wild: Insights from Social Carnivores." *Philosophical Transactions of the Royal Society B: Biological Sciences* 373, no. 1740 (February 19, 2018): 20160508.

Berman, Russell A. *Fiction Sets You Free: Literature, Liberty, and Western Culture*. Iowa City: University of Iowa Press, 2007.

Berggren, J. Lennart. "Medieval Arithmetic: Arabic Texts and European Motivations." In *Word, Image, Number: Communication in the Middle Ages*. Edited by John J. Contreni and Santa Casciani. Tavarnuzze (Florence), Italy: SISMEL Edizioni del Galluzzo, 2002,

Biagioli, Mario. *Galileo Courtier*. Chicago: University of Chicago Press, 1993.

Biden, Joseph R. Jr. "A Proclamation on the Public Service Recognition Week, 2022." April 29, 2022. https://www.whitehouse.gov/briefing-room/presidential-actions/2022/04/29/a-proclamation-on-public-service-recognition-week-2022/.

Biggs, Norman. "Mathematics of Currency and Exchange: Arithmetic at the End of the Thirteenth Century." *British Society for the History of Mathematics Bulletin* 24 (2009): 67–77.

Blair, Ann. "Note Taking as an Art of Transmission." *Critical Inquiry* 31, no. 1 (Autumn 2004): 85–107.

Blair, Ann. *Too Much to Know: Managing Scholarly Information before the Modern Age*. New Haven, CT: Yale University Press, 2010.

Bloom, Gina. *Gaming the Stage: Playable Media and the Rise of English Commercial Theatre*. Ann Arbor: University of Michigan Press, 2018.

Boulton, Jeremy, and Leonard Schwarz. "Yet Another Inquiry into the Trustworthiness of Eighteen-century London's Bills of Mortality." *Local Population Studies* 85 (2010): 28–45.

Borst, Arno. *The Ordering of Time: From the Ancient Computus to the Modern Computer*. Translated by Andrew Winnard. Chicago: University of Chicago Press, 1993.

Braddick, Michael J. *State Formation in Early Modern England, 1550–1700*. Cambridge: Cambridge University Press, 2000.

Brearley, Harry C. *Time Telling through the Ages*. New York: Doubleday, Page, 1919.

Bregman, Alvan. "Alligation Alternate and the Composition of Medicines: Arithmetic and Medicine in Early Modern England." *Medical History* 49 (2005): 299–320.

Breisach, Ernst. *Historiography: Ancient, Medieval, and Modern*, 2nd ed. Chicago: University of Chicago Press, 1994.

Bruce, Scott G. *Silence and Sign Language in Medieval Monasticism: The Cluniac Tradition, c. 900–1200*. Cambridge: Cambridge University Press, 2010.

Burdick, Bruce Stanley. *Mathematical Words Printed in the Americas, 1554–1700*. Baltimore: Johns Hopkins University Press, 2009.

Brooks, Colin. "Projecting, Political Arithmetic and the Act of 1695." *English Historical Review* 97, no. 382 (January 1982): 31–53.

Buck, Peter. "People Who Counted: Political Arithmetic in the Eighteenth Century." *Isis* 73, no. 1 (March 1982): 28–45.

Buck, Peter. "Seventeenth-Century Political Arithmetic: Civil Strife and Vital Statistics." *Isis* 68, no. 1 (March 1977): 67–84.

Burke, Peter. *A Social History of Knowledge from Gutenberg to Diderot*. Cambridge: Polity, 2000.

Burke, Peter. *Popular Culture in Early Modern Europe*. New York: New York Press, 1978.

Burke, Peter. "Popular Culture in Seventeenth-Century London." In *Popular Culture in Seventeenth-Century England*. Edited by Barry Reay, 31–58. London: Croom Helm, 1985.

Burke, Peter. "The Renaissance Dialogue." *Renaissance Studies* 3, no. 1 (March 1989): 1–12.

Burke, Peter. *Varieties of Cultural History*. Cambridge: Polity Press, 1997.

Burr, David C., Giovanni Anobile, and Roberto Arrighi. "Psychophysical Evidence for the Number Sense." *Philosophical Transactions of the Royal Society B: Biological Sciences* 373, no. 1740 (February 19, 2018): 20170045.

Butler, Christopher. *Number Symbolism*. New York: Barnes and Noble, 1970.

Butterfield, Herbert. *The Origins of Modern Science, 1300–1800*. New York: Macmillan, 1951.

Butterworth, Brian. "The Implications for Education of an Innate Numerosity-Processing Mechanism." *Philosophical Transactions of the Royal Society B: Biological Sciences* 373, no. 1740 (February 19, 2018): 20170118.

Butterworth, Brian. *What Counts: How Every Brain Is Hardwired for Math*. New York: Free Press, 1999.

Butterworth, Brian, C. R. Gallistel, and Giorgio Vallortigara. "Introduction: The Origins of Numerical Abilities." *Philosophical Transactions of the Royal Society B: Biological Sciences* 373, no. 1740 (February 19, 2018): 20160507.

Butterworth, Brian, and Robert Reeve. "Verbal Counting and Spatial Strategies in Numerical Tasks: Evidence from Indigenous Australia." *Philosophical Psychology* 21, no. 4 (August 1, 2008): 443–57.

Butterworth, Brian, Sashank Varma, and Diana Laurillard. "Dyscalculia: From Brain to Education." In *The Oxford Handbook of Numerical Cognition*. Edited by Roi Cohen Kadosh and Ann Dowker, 647–61. Oxford: Oxford University Press, 2015.

Cahill, Patricia. "Killing by Computation: Military Mathematics, the Elizabethan Social Body, and Marlowe's *Tamburlaine*." In *Arts of Calculation: Quantifying Thought in Early Modern Europe*. Edited by David Glimp and Michelle R. Warren, 165–86. New York: Palgrave Macmillan, 2004.

Cajori, Florian. *A History of Elementary Mathematics*. New York: Macmillan Company, 1921.

Cajori, Florian. *A History of Mathematical Notations*. Chicago: Open Court, 1929.

Cambers, Andrew. "Demonic Possession, Literacy and 'Superstition' in Early Modern England." *Past and Present* 202 (2009): 3–35.

Campi, Emidio, Simone De Angelis, Anja-Silvia Goeing, and Anthony T. Grafton, eds. *Scholarly Knowledge: Textbooks in Early Modern Europe*. Geneva, Switzerland: Librairie Droz, 2008.

Capp, Bernard. *Astrology and the Popular Press: English Almanacs 1500–1800*. Ithaca, NY: Cornell University Press, 1979.

Capp, Bernard. "Popular Literature." In *Popular Culture in Seventeenth-Century England*. Edited by Barry Reay, 198–243. London: Croom Helm, 1985.

Carey, James W. "The Paradox of the Book." *Library Trends* 33, no. 2 (Fall, 1984): 103–113.

Carpenter, David A. *The Struggle for Mastery: Britain, 1066–1284*. London: Penguin Books, 2005.

Cassedy, James H. *Demography in Early America: Beginnings of the Statistical Mind, 1600–1800*. Cambridge, MA: Harvard University Press, 1969.

Chapman, Alison A. "Marking Time: Astrology, Almanacs, and English Protestantism." *Renaissance Quarterly* 60 (2007): 1257–90.

Chapman, Alison A. "The Politics of Time in Edmund Spenser's English Calendar." *Studies in English Literature, 15001900* 42, no. 1 (2002): 1–24.

Charlton, Kenneth. *Education in Renaissance England*. London: Routledge and Kegan Paul, 1965.

Charlton, Kenneth. *Women, Religion and Education in Early Modern England*. London: Routledge, 1999.

Chrisomalis, Stephen. *Numerical Notation: A Comparative History*. Cambridge: Cambridge University Press, 2010.

Chrisomalis, Stephen. *Reckonings: Numerals, Cognition, and History*. Cambridge, MA: MIT Press, 2020.

Chrisomalis, Stephen. "The Cognitive and Cultural Foundations of Numbers." In *The Oxford Handbook of the History of Mathematics*. Edited by Eleanor Robson and Jacqueline Stedall, 495–518. Oxford: Oxford University Press, 2009.

Chrisomalis, Stephen. "The Origins and Co-Evolution of Literacy and Numeracy." In *The Cambridge Handbook of Literacy*. Edited by David R. Olson and Nancy Torrance, 59–74. Cambridge: Cambridge University Press, 2009.

Cipolla, Carlo M. *Before the Industrial Revolution: European Society and Economy, 1000–1700*. London: Methuen, 1976.

Cipolla, Carlo M. *Clocks and Culture, 1300–1700*. New York: W.W. Norton, 1978.

Clanchy, M. T. *From Memory to Written Record: England 1066–1307*, 3rd ed. Oxford: Wiley-Blackwell, 2013.

Clark, Geoffrey. *Betting on Lives: The Culture of Life Insurance in England, 1695–1775*. Manchester: Manchester University Press, 1999.

Clark, William, Jan Golinski, and Simon Schaffer, eds. *The Sciences in Enlightened Europe*. Chicago: University of Chicago Press, 1999.

Clawson, Calvin C. *The Mathematical Traveler: Exploring the Grand History of Numbers*. New York: Plenum Press, 1994.

Cobbett, William. *The Parliamentary History of England, from the Earliest Period to the Year 1803*, vol. 23. London: R. Bagshaw, 1814.

Cohen, Patricia Cline. *A Calculating People: The Spread of Numeracy in Early America*. Chicago: University of Chicago Press, 1982.

Collinson, Patrick. *The Elizabethan Puritan Movement*. Oxford: Clarendon Press, 1967.

Corfield, Penelope J. *Time and the Shape of History*. New Haven, CT: Yale University Press, 2007.

Cormack, Bradin, and Carla Mazzio. *Book Use, Book Theory: 1500–1700*. Chicago: University of Chicago Press, 2005.

Cressy, David. *Birth, Marriage and Death: Ritual, Religion, and the Life-Cycle in Tudor and Stuart England*. Oxford: Oxford University Press, 1997.

Cressy, David. *Bonfires and Bells: National Memory and the Protestant Calendar in Elizabethan and Stuart England*. London: Weidenfeld and Nicolson, 1989.

Cressy, David. *Education in Tudor and Stuart England*. New York: St. Martin's Press, 1975.

Cressy, David. *Literacy and the Social Order: Reading and Writing in Tudor and Stuart England*. Cambridge: Cambridge University Press, 1980.

Cressy, David. "The Protestant Calendar and the Vocabulary of Celebration in Early Modern England." *Journal of British Studies* 29, no. 1 (January 1990): 31–52.

Crombie, A. C. *Robert Grosseteste and the Origins of Experimental Science, 1100–1700*. Oxford: Clarendon Press, 1953.

Crosby, Alfred W. *The Measure of Reality: Quantification and Western Society, 1250–1600*. Cambridge: Cambridge University Press, 1997.

Cruz, Helen De. "An Extended Mind Perspective on Natural Number Representation." *Philosophical Psychology* 21, no. 4 (August 1, 2008): 475–90.

Cunningham, Hugh. *The Children of the Poor: Representations of Childhood since the Seventeenth Century*. Oxford: Basil Blackwell, 1991.

Curry, Patrick. *Prophecy and Power: Astrology in Early Modern England.* Princeton, NJ: Princeton University Press, 1989.

Curry, Patrick. "Astrology in Early Modern England: The Making of Vulgar Knowledge." In *Science, Culture and Popular Belief in Renaissance Europe.* Edited by Stephen Pumfrey, Paolo L. Rossi, and Maurice Slawinski, 274–92. Manchester: Manchester University Press, 1991.

Curtis, Mark H. *Oxford and Cambridge in Transition: 1558–1642.* Oxford: Clarendon Press, 1959.

Dane, Joseph A. *The Myth of Print Culture: Essays on Evidence, Textuality, and Bibliographical Method.* Toronto: University of Toronto Press, 2003.

Dane, Joseph A., and Alexandra Gillespie. "The Myth of the Cheap Quarto." In *Tudor Books and Readers: Materiality and the Construction of Meaning.* Edited by John N. King, 25–45. Cambridge, 2010.

Daston, Lorraine. *Classical Probability in the Enlightenment.* Princeton, NJ: Princeton University Press, 1998.

David, F. N. *Games, Gods and Gambling: The Origins and History of Probability and Statistical Ideas from the Earliest Times to the Newtonian Era.* New York: Hafner, 1962.

Davies, Margaret Gay. *The Enforcement of English Apprenticeship: A Study in Applied Mercantilism, 1563–1642.* Cambridge, MA: Harvard University Press, 1956.

De Cruz, Helen, and Johan De Smedt. "Mathematical Symbols as Epistemic Actions." *Synthese* 190, no. 1 (January 1, 2013): 3–19.

De Smedt, John, and Helen De Cruz. " The Role of Material Culture in Human Time Representation: Calendrical Systems as Extensions of Mental Time Travel." *Adaptive Behavior* 19, no. 1 (2011): 63–76.

Dean, David. "Elizabeth's Lottery: Political Culture and State Formation in Early Modern England." *Journal of British Studies* 50, no. 3 (2011): 587–611.

Dehaene, Stanislas. *The Number Sense: How the Mind Creates Mathematics.* New York: Oxford University Press, 1997.

Dehaene, Stanislas, and Laurent Cohen. "Cultural Recycling of Cortical Maps." *Neuron* 56, no. 2 (October 25, 2007): 384–98.

Dehaene, Stanislas, Laurent Cohen, José Morais, and Régine Kolinsky. "Illiterate to Literate: Behavioural and Cerebral Changes Induced by Reading Acquisition." *Nature Reviews Neuroscience* 16, no. 4 (April 2015): 234–44.

De Morgan, Augustus. *Arithmetical Books from the Invention of Printing to the Present Time Being Brief Notices of a Large Number of Works Drawn Up from Actual Inspection.* London: Tayor and Walton, 1847.

DeMolen, Richard L. *Richard Mulcaster and Educational Reform in the Renaissance.* Nieuwkoop: De Graaf, 1991.

Denniss, John. "Arithmetical Textbooks 1478–1886: A Progression?" *British Society for the History of Mathematics Bulletin* 21 (2006): 26–33.

Denniss, John. *Figuring It Out: Children's Arithmetical Manuscripts, 1680–1880.* Oxford: Huxley Scientific Press, 2012.

Denniss, John. "Learning Arithmetic: Textbooks and Their Users in England 1500–1900." In *The Oxford Handbook of the History of Mathematics.* Edited by Eleanor Robson and Jacqueline Stedall, 448–67. Oxford: Oxford University Press, 2009.

Deringer, William. *Calculated Values: Finance, Politics, and the Quantitative Age.* Cambridge, MA: Harvard University Press, 2018.

Desan, Christine. *Making Money: Coin, Currency, and the Coming of Capitalism*. Oxford: Oxford University Press, 2015.

Desborough, Jane. *The Changing Face of Early Modern Time, 1550–1770*. Cham, Switzerland: Palgrave Macmillan, 2019.

Desoete, Annemie. "Cognitive Predictors of Mathematical Abilities and Disabilities." In *The Oxford Handbook of Numerical Cognition*. Edited by Roi Cohen Kadosh and Ann Dowker, 915–32. Oxford: Oxford University Press, 2015.

Devlin, Keith. *The Unfinished Game: Pascal, Fermat, and the Seventeenth-Century Letter that Made the World Modern*. New York: Basic Books, 2008.

Dickinson, Michael. *Seventeenth Century Tokens of the British Isles and Their Values*. London: Seaby, 1986.

Dobbs, Bety Jo Teeter. *The Janus Face of Genius: The Role of Alchemy in Newton's Thought*. Cambridge: Cambridge University Press, 1991.

Dohrn-van Rossum, Gerhard. *History of the Hour: Clocks and Modern Temporal Orders*. Translated by Thomas Dunlap. Chicago: University of Chicago Press, 1996.

Dotan, Dror, and Stanislas Dehaene. "On the Origins of Logarithmic Number-to-Position Mapping." *Psychological Review* 123, no. 6 (20161031): 637.

Dover, Paul M. *The Information Revolution in Early Modern Europe*. Cambridge: Cambridge University Press, 2021.

Duhem, Pierre. *Le Système Du Monde: Histoire Des Doctrines Cosmologiques de Platon à Copernic*. Paris: A. Hermann, 1913.

Ebert, Christopher. "Early Modern Atlantic Trade and the Development of Maritime Insurance to 1630." *Past and Present* 213 (2011): 87–114.

Eisenstein, Elizabeth. *The Printing Press as an Agent of Change*. Cambridge: Cambridge University Press, 1979.

Ellerton, Nerida, and M. A (Ken) Clements. *Rewriting the History of School Mathematics in North America, 1607–1861: The Central Role of Cyphering Books*. Dordrecht: Springer, 2012.

Elton, G. R. *The Tudor Revolution in Government: A Study of Administrative Changes in the Reign of Henry VIII*. Cambridge: Cambridge University Press, 1969.

Ernest, Paul. "A Semiotic Perspective of Mathematical Activity: The Case of Number." *Educational Studies in Mathematics* 61, no. 1/2, Semiotic Perspectives in Mathematics Education: A PME Special Issue (2006): 67–101.

Errico, Francesco d', Luc Doyon, Ivan Colagé, Alain Queffelec, Emma Le Vraux, Giacomo Giacobini, Bernard Vandermeersch, and Bruno Maureille. "From Number Sense to Number Symbols. An Archaeological Perspective." *Philosophical Transactions of the Royal Society B: Biological Sciences* 373, no. 1740 (February 19, 2018): 20160518.

Everett, Caleb. *Numbers and the Making of Us: Counting and the Course of Human Cultures*. Cambridge, MA: Harvard University Press, 2017.

Fauvel, John, and Robert Goulding. "Renaissance Oxford." In *Oxford Figures: 800 Years of the Mathematical Sciences*. Edited by John Fauvel, Raymond Flood, and Robin Wilson, 41–62. Oxford: Oxford University Press, 2000.

Fauvel, John, and Jeremy Gray, eds. *The History of Mathematics: A Reader*. New York: Palgrave Macmillan, 1987.

Febvre, Lucien, and Henri-Jean Martin. *The Coming of the Book: The Impact of Printing 145--1800*. Translated by David Gerard. Edited by Geoffrey Nowell-Smith and David Wootton. London: Verso Edition, 1984.

Feingold, Mordechai. "Decline and Fall: Arabic Science in Seventeenth-Century England." In *Tradition, Transmission, Transformation: Proceedings of Two Conferences on Pre-Modern Science Held at the University of Oklahoma*. Edited by F. Jamil Ragep and Sally Ragep, 441–69. Leiden: E. J. Brill, 1996.

Feingold, Mordechai. *The Mathematicians' Apprenticeship: Science, Universities and Society in England, 1560–1640*. Cambridge: Cambridge University Press, 1984.

Feingold, Mordechai. "Reading Mathematics in the English Collegiate-Humanist Universities." In *Reading Mathematics in Early Modern Europe: Studies in the Production, Collection, and Use of Mathematical Books*. Edited by Philip Beeley, Yelda Nasifoglu, and Benjamin Wardhaugh, 124–50. New York: Routledge, 2021.

Feingold, Mordechai. "Tradition versus Novelty: Universities and Scientific Studies in the Early Modern Period." In *Revolution and Continuity Essays in the History and Philosophy of Early Modern Science*. Edited by Peter Barker and Roger Ariew, 45–59. Washington, DC: Catholic University of America Press, 1991.

Feist, Timothy. *The Stationers' Voice: The English Almanac Trade in the Early Eighteenth Century*. Philadelphia: American Philosophical Society, 2005.

Felski, Rita. "The Invention of Everyday Life." In *New Formations* 39 (1999): 15–31.

Field, Jacob F. *London, Londoners and the Great Fire of 1666: Disaster and Recovery*. New York: Routledge, 2018.

Flegg, Graham. *Numbers: Their History and Meaning*. New York: Schocken Books, 1983.

Fletcher, C. R. L., ed. *Collectanea: First Series*. Oxford: Clarendon Press, 1885.

Fox, Adam. *Oral and Literate Culture in England 1500–1700*. Oxford: Clarendon Press, 2000.

Franklin, James. *The Science of Conjecture: Evidence and Probability before Pascal*. Baltimore: Johns Hopkins University Press, 2001.

Frye, Douglass, Nicholas Braisby, John Lowe, Celine Maroudas, and Jon Nicholls. "Young Children's Understanding of Counting and Cardinality." *Child Development* 60, no. 5 (October 1989): 1158–71.

Gagné, John. "Counting the Dead: Traditions of Enumeration and the Italian Wars." *Renaissance Quarterly* 67 (2014): 791–849.

Gardiner, Samuel Rawson. *The Constitutional Documents of the Puritan Revolution 1625–1600*. Oxford: Clarendon Press, 1906.

Geary, David C. "The Classification and Cognitive Characteristics of Mathematical Disabilities in Children." In *The Oxford Handbook of Numerical Cognition*. Edited by Roi Cohen Kadosh and Ann Dowker, 767–86. Oxford: Oxford University Press, 2015.

Gebuis, Titia and Bert Reynvoet. "Number Representations and Their Relation with Mathematical Ability." In *The Oxford Handbook of Numerical Cognition*. Edited by Roi Cohen Kadosh and Ann Dowker, 331–44. Oxford: Oxford University Press, 2015.

Geertz, Clifford. *The Interpretation of Cultures: Selected Essays*. New York: Basic Books, 1973.

Geertz, Clifford. *Local Knowledge: Further Essays in Interpretive Anthropology*. New York: Basic Books, 1983.

Geneva, Ann. *Astrology and the Seventeenth Century Mind: William Lilly and the Language of the Stars*. Manchester: Manchester University Press, 1995.

Gerdes, Paulus. *Geometry from Africa: Mathematical and Educational Explorations*. Washington, DC: Mathematical Association of America, 1999.

Gigerenzer, Gerd. *Empire of Chance: How Probability Changed Science and Everyday Life*. Cambridge: Cambridge University Press, 1989.

Gingerich, Owen. *The Book Nobody Read: Chasing the Revolutions of Nicolaus Copernicus.* New York: Walker, 2004.

Glaisyer, Natash, *The Culture of Commerce in England, 1660–1720.* Woodbridge, Suffolk: Boydell Press, 2006.

Glaisyer, Natasha. "Popular Didactic Literature." In *The Oxford History of Popular Print Culture*, vol. 1, *Cheap Print in Britain and Ireland to 1660.* Edited by Joad Raymond, 1:510–9. Oxford: Oxford University Press, 2011.

Glaisyer, Natasha, and Sara Pennell, eds. *Didactic Literature in England, 1500–1800: Expertise Constructed.* Aldershot, Hampshire, England: Ashgate, 2003.

Glass, D. V. "Gregory King's Estimate of the Population of England and Wales, 1695." *Population Studies* 3, no. 4 (March 1950): 338–74.

Glennie, Paul, and Nigel Thrift. *Shaping the Day: A History of Timekeeping in England and Wales, 1300–1800.* Oxford: Oxford University Press, 2009.

Glimp, David, and Michelle R. Warren. *Arts of Calculation: Quantifying Thought in Early Modern Europe.* New York: Palgrave Macmillan, 2004.

Golinski, Jan. *Making Natural Knowledge: Constructivism and the History of Science.* Cambridge: Cambridge University Press, 1998.

Golinski, Jan. "The Theory of Practice and the Practice of Theory: Sociological Approaches in the History of Science." *Isis* 81, no. 3 (September 1990): 492–505.

Grafton, Anthony. *Bring Out Your Dead: The Past as Revelation.* Cambridge, MA: Harvard University Press, 2001.

Gray, Jeremy. "History of Mathematics and History of Science Reunited?" *Isis* 102 (2011): 511–7.

Green, Ian. *The Christian's ABC: Catechisms and Catechizing in England, c.1530–1740.* Oxford: Clarendon Press, 1996.

Green, Ian. *Humanism and Protestantism in Early Modern English Education.* Burlington, VT: Ashgate, 2009.

Green, Monica H. "The Four Black Deaths." *American Historical Review* 125, no. 5 (2020): 1601–31.

Green, Monica H. "Putting Africa on the Black Death Map: Narratives from Genetics and History." *Afriques* 9 (2018). https://doi.org/10.4000/afriques.2125.

Greenberg, Stephen. "Plague, Printing, and Public Health." *Huntington Library Quarterly* 67, no. 4 (December 2004): 508–27.

Griffiths, Paul. "Local Arithmetic: Information Cultures in Early Modern England." In *Remaking English Society: Social Relations and Social Change in Early Modern England.* Edited by Steve Hindle, Alexandra Shepard, and John Walter, 113–34. Woodbridge, Suffolk: Boydell Press, 2013.

Griffiths, Paul. "Surveying the People." In *A Social History of England: 1500–1750.* Edited by Keith Wrightson, 39–59. Cambridge: Cambridge University Press, 2017.

Groza, Vivian Shaw. *A Survey of Mathematics: Elementary Concepts and Their Historical Development.* New York: Holt, Rinehart and Winston, 1968.

Guibbory, Achsah. *The Map of Time: Seventeenth-Century English Literature and Ideas of Pattern in History.* Urbana: University of Illinois Press, 1986.

Gunther, R. T. *Early Science in Oxford.* Oxford: Clarendon Press, 1920-68.

Gvozdanovic, Jadranka, ed. *Numeral Types and Changes Worldwide.* New York: Mouton de Grutyer, 1999.

Hackel, Heidi Brayman. "Popular Literacy and Society." In *The Oxford History of Popular Print Culture*, vol. 1, *Cheap Print in Britain and Ireland to 1660*. Edited by Joad Raymond, 88–100. Oxford: Oxford University Press, 2011.

Hacking, Ian. *The Emergence of Probability: A Philosophical Study of Early Ideas about Probability Induction and Statistical Inference*, 2nd ed. Cambridge: Cambridge University Press, 2006.

Hacking, Ian. *The Social Construction of What?* Cambridge, MA: Harvard University Press, 1999.

Hadden, Richard W. *On the Shoulders of Merchants: Exchange and the Mathematical Conception of Nature in Early Modern Europe*. Albany: State University of New York Press, 1994.

Hailwood, Mark. "Time and Work in Rural England, 1500--700." *Past and Present*, 248, no. 1 (2020): 87–121.

Hall, A. Rupert. *The Scientific Revolution, 1500–1800: The Formation of the Modern Scientific Attitude*. New York: Longmans, Green, 1954.

Hanawalt, Barbara A. *Growing up in Medieval London: The Experience of Childhood in History*. New York: Oxford University Press, 1993.

Hannagan, Thomas, Amir Amedi, Laurent Cohen, Ghislaine Dehaene-Lambertz, and Stanislas Dehaene. "Origins of the Specialization for Letters and Numbers in Ventral Occipitotemporal Cortex." *Trends in Cognitive Sciences* 19, no. 7 (July 1, 2015): 374–82.

Harkness, Deborah E. "A View from the Streets: Women and Medical Work in Elizabethan London." *Bulletin of the History of Medicine* 82, no. 1 (2008): 52–85.

Harkness, Deborah E. *The Jewel House: Elizabethan London and the Scientific Revolution*. New Haven, CT: Yale University Press, 2007.

Harkness, Deborah E. "Managing an Experimental Household: The Dees of Mortlake and the Practice of Natural Philosophy." *Isis* 88, no. 2 (June 1997): 247–62.

Hargrave, Catherine Perry. *A History of Playing Cards and a Bibliography of Cards and Gaming*. Boston: Houghton Mifflin, 1980.

Harrington, J. C. "Evidence of Manual Reckoning in The Cittie of Ralegh." *North Carolina Historical Review* 33 no. 1 (January 1956): 1–11.

Harvey, Karen, ed. *History and Material Culture: A Student's Guide to Approaching Alternative Sources*. New York: Routledge, Taylor & Francis, 2009.

Hawkins, Edward, compiler; Augustus W. Franks and Herbert A. Grueber, eds. *Medallic Illustrations of the History of Great Britain and Ireland to the Death of George II*. London: Spink and Son, 1969.

Hawkes, David. *The Culture of Usury in Renaissance England*. New York: Palgrave Macmillan, 2010.

Headrick, Daniel R. *When Information Came of Age: Technologies of Knowledge in the Age of Reason and Revolution*. New York: Oxford University Press, 2000.

Hebra, Alex. *Measure for Measure: The Story of Imperial, Metric, and Other Units*. Baltimore: Johns Hopkins University Press, 2003.

Heitman, Kristin. "Authority, Autonomy and the First London Bills of Mortality." *Centaurus* 62, no. 2 (2020), 278.

Henry, Wanda S. "Women Searchers of the Dead in Eighteenth- and Nineteenth-Century London." *Social History of Medicine* 29, no. 3 (2016): 445–66.

Herman, Bernard L. *The Stolen House*. Charlottesville: University of Virginia Press, 1992.

Higgs, Edward. *The Information State in England*. New York: Palgrave Mcmillan, 2004.

Hill Curth, Louise. *English Almanacs, Astrology and Popular Medicine: 1550-1700.* Manchester: Manchester University Press, 2007.

Hill, G. F. *The Development of Arabic Numerals in Europe: Exhibited in 64 Tables.* Oxford: Clarendon Press, 1915.

Hindle, Steve. *The State and Social Change in Early Modern England c. 1550-1640.* New York: St. Martin's Press, 2000.

Hobart, Michael E., and Zachary S. Schiffman. *Information Ages: Literacy, Numeracy, and the Computer Revolution.* Baltimore: Johns Hopkins University Press, 1998.

Holloway, Ian D. and Daniel Ansari. "Numerical Symbols: An Overview of Their Cognitive and Neural Underpinnings." In *The Oxford Handbook of Numerical Cognition.* Edited by Roi Cohen Kadosh and Ann Dowker, 531–51. Oxford: Oxford University Press, 2015.

Holmes, T. Rice. "The Birthday of Augustus and the Julian Calendar." *Classical Quarterly* 6, no. 2 (April 1912): 73–81.

Holt, Peter Malcolm. *Studies in the History of the Near East.* London: Frank Cass, 1973.

Hoppit, Julian. "Political Arithmetic in Eighteenth-Century England." *Economic History Review,* n.s., 49, no. 3 (1996): 516–40.

Horrocks, Thomas A. *Popular Print and Popular Medicine: Almanacs and Health Advice in Early America.* Amherst: University of Massachusetts Press, 2008.

Howard, Geoffrey. *A History of Mathematics Education in England.* Cambridge: Cambridge University Press, 1982.

Hoyle, R. W. "Crown, Parliament and Taxation in Sixteenth-Century England." *English Historical Review* 109, no. 434 (November 1994): 1174–96.

Hubbard, Eleanor. "Reading, Writing, and Initialing: Female Literacy in Early Modern London." *Journal of British Studies* 54, no. 3 (July 2015): 553–77.

Huff, Toby E. *The Rise of Early Modern Science: Islam, China, and the West.* Cambridge: Cambridge University Press, 2003.

Hughes, Barnabas. "Robert Recorde and the First Published Equation." In *Vestigia Mathematica: Studies in Medieval and Early Modern Mathematics in Honour of H.L.L. Busard,* Edited by M. Folkerts and Jan Hogendijk, 163–71. Amsterdam: Rodopi B. V., 1993.

Huizinga, Johan. *Homo Ludens: A Study of the Play Element in Culture.* Boston: Beacon Press, 1950.

Hurford, James R. *Language and Number: The Emergence of a Cognitive System.* Oxford: Basil Blackwell, 1987.

Hutton, Ronald. *The Stations of the Sun: A History of the Ritual Year in Britain.* Oxford: Oxford University Press, 1996.

Ifrah, Georges. *From One to Zero: A Universal History of Numbers.* Translated by Lowell Bair. New York: Viking Penguin, 1985.

Jackson, T. W. "Dr. Wallis' Letter against Mr. Maidwell, 1700." In *Collectanea,* 1st ser. Edited by C. R. L. Fletcher, 269–337. Oxford: Clarendon Press, 1885.

Jacobs Danan, Jennifer A., and Rochel Gelman. "The Problem with Percentages." *Philosophical Transactions of the Royal Society B: Biological Sciences* 373, no. 1740 (February 19, 2018): 20160519.

James, Kathryn. "Reading Numbers in Early Modern England." *British Society for the History of Mathematics Bulletin* 26 (2011): 1–16.

James, Susan E. *Women's Voices in Tudor Willis, 1485-1603.* New York: Routledge, 2016.

Jenkins, David, and Takau Yoneyama, eds. *History of Insurance*. London: Pickering and Chatto, 2000.

Jenkinson, Hilary. "An Original Exchequer Account of 1304 with Private Tallies Attached." In *Proceedings of the Society of Antiquaries of London, 27th November 1913 to 25th June 1914*, 2nd ser., vol. 26 by Society of Antiquaries of London. Oxford: Horace Hart, 1914.

Jenkinson, Hilary. "Exchequer Tallies." *Archaeologia or, Miscellaneous tracts relating to antiquity*, 2nd ser., 62 (1911): 367–80.

Jenkinson, Hilary. "Medieval Tallies, Public and Private." In *Selected Writings of Sir Hilary Jenkinson*. Gloucester, England: Alan Sutton, 1980.

Jewell, Helen M. *Education in Early Modern England*. New York: St. Martin's Press, 1998.

Johns, Adrian. *The Nature of the Book: Print and Knowledge in the Making*. Chicago: University of Chicago Press, 1998.

Johns, Adrian. "Science and the Book." In *The Cambridge History of the Book in Britain, vol. 4, 1557–1695*. Edited by John Barnard and D. F. McKenzie, 274–303. Cambridge: Cambridge University Press, 2002.

Johnson, Harry M. "The History of British and American Fire Marks." *Journal of Risk and Insurance* 39, no. 3 (September 1972): 405–18.

Johnston Gordon, Rona. "Controlling Time in the Hapsburg Lands: The Introduction of the Gregorian Calendar in Austria Below the Enns." *Austrian History Yearbook* 40 (2009): 28–36.

Jones, Norman. *God and the Moneylenders*. Oxford: Basil Blackwell, 1989.

Joyce, Patrick. "What Is the Social in Social History?" *Past and Present* 206 (February 2010): 213–48

Jutte, Robert. *A History of the Senses: from Antiquity to Cyberspace*. Cambridge: Polity, 2005.

Kaiser, David, ed. *Pedagogy and the Practice of Science: Historical and Contemporary Perspectives*. Cambridge, MA: MIT Press, 2005.

Kaufmann, Miranda. *Black Tudors: The Untold Story*. London: Oneworld, 2017.

Kavanagh, Thomas M. *Dice, Cards, Wheels: A Different History of French Culture*. Philadelphia: University of Pennsylvania Press, 2005.

Kelly, Thomas. *A History of Adult Education in Great Britain*. Liverpool: Liverpool University Press, 1992.

Kerridge, Eric. *Trade and Banking in Early Modern England*. Manchester: Manchester University Press, 1988.

Koyré, Alexander. *From the Closed World to the Infinite Universe*. Baltimore: Johns Hopkins Press, 1957.

Kunitzsch, Paul. "The Transmission of Hindu-Arabic Numerals Reconsidered." In *The Enterprise of Science in Islam: New Perspectives*. Edited by Jan Hogendijk and Abdelhamid I. Sabra, 3–22. Cambridge: MIT Press, 2003.

Lagomarsino, David, and Charles T. Wood, eds. *The Trial of Charles I: A Documentary History*. Hanover, NH: University Press of New England, 1989.

Lake Prescott, Anne. "Refusing Translation: The Gregorian Calendar and Early Modern English Writers." *Yearbook of English Studies* 36, no. 1 (2006): 1–11.

Landes, David S. *Revolution in Time: Clocks and the Making of the Modern World*. London: Penguin, 2000.

Lane, Joan. *Apprenticeship in England: 1600–1914*. Boulder, CO: Westview Press, 1996.

Lawson, John, and Harold Silver. *A Social History of Education in England*. London: Routledge, 1973.

Lee, G. A. "The Oldest European Account Book: A Florentine Bank Ledger of 1211." In *Accounting History: Some British Contributions*. Edited by R. H. Parker and B. S. Yamey, 116–38. Oxford: Oxford University Press, 1994.

LeGoff, Jacques. *Time, Work, and Culture in the Middle Ages*. Translated by Arthur Goldhammer. Chicago: University of Chicago Press, 1980.

Lefevre, Jo-Anne, Emma Wells, and Carla Sowinski. "Individual Differences in Basic Arithmetical Processes in Children and Adults." In *The Oxford Handbook of Numerical Cognition*. Edited by Roi Cohen Kadosh and Ann Dowker, 895–914. Oxford: Oxford University Press, 2015.

Lindberg, David C. *The Beginnings of Western Science: The European Scientific Tradition in Philosophical, Religious, and Institutional Context, 600 B.C. to A.D. 1450*. Chicago: University of Chicago Press, 1992.

Lindberg, David C., and Robert Westman. *Reappraisals of the Scientific Revolution*. Cambridge: Cambridge University Press, 1990.

Lindemann, Oliver and Martin H. Fischer. "Cognitive Foundations of Human Number Representations and Mental Arithmetic." In *The Oxford Handbook of Numerical Cognition*. Edited by Roi Cohen Kadosh and Ann Dowker, 35–44. Oxford: Oxford University Press, 2015.

Littleton, Ananias Charles, and Basil S. Yamey. *Studies in the History of Accounting*. New York: Arno Press, 1978.

Lockhart, Paul. *Arithmetic*. Cambridge, MA: Harvard University Press, 2017.

Lopez, Robert S. *The Commercial Revolution of the Middle Ages, 950–1350*. Englewood Cliffs, NJ: Prentice-Hall, 1971.

López-Barroso, Diana, Michel Thiebaut de Schotten, José Morais, Régine Kolinsky, Lucía W. Braga, Alexandre Guerreiro-Tauil, Stanislas Dehaene, and Laurent Cohen. "The Impact of Early and Late Literacy on the Functional Connectivity of Vision and Language-Related Networks." *NeuroImage* 213 (2020).

Lux, David S. "Societies, Circles, Academies, and Organizations: A Historiographic Essay on Seventeenth-Century Science." In *Revolution and Continuity Essays in the History and Philosophy of Early Modern Science*. Edited by Peter Barker and Roger Ariew, 23–43. Washington, DC: Catholic University of America Press, 1991.

Macey, Samuel L. *Clocks and the Cosmos: Time in Western Life and Thought*. Hamden, CT: Archon Books, 1980.

Maltby, Judith. *Prayer Book and People in Elizabethan and Early Stuart England*. Cambridge: Cambridge University Press, 1998.

Mann, Tony. "History of Mathematics and History of Science." *Isis* 102 (2011): 518–26.

Mason, John. "Bartering Problems in Arithmetic Books 1450–1890." *British Society for the History of Mathematics Bulletin* 22 (2007): 160–81.

Mazzocco, Michèle M. M. "The Contributions of Syndrome Research to the Study of MLD." In *The Oxford Handbook of Numerical Cognition*. Edited by Roi Cohen Kadosh and Ann Dowker, 678–95. Oxford: Oxford University Press, 2015.

McCormick, Ted. *Human Empire: Mobility and Demographic Thought in the British Atlantic World, 1500–1800*. Cambridge: Cambridge University Press, 2022.

McCormick, Ted. *William Petty and the Ambitions of Political Arithmetic*. Oxford: Oxford University Press, 2009.

McCrink, Koleen and Wesley Birdsall. "Numerical Abilities and Arithmetic in Infancy." In *The Oxford Handbook of Numerical Cognition*. Edited by Roi Cohen Kadosh and Ann Dowker, 258–74. Oxford: Oxford University Press, 2015.

McCrink, Koleen, and Karen Wynn. "Large-Number Addition and Subtraction by 9-Month-Old Infants." *Psychological Science* 15, no. 11 (November 2004): 776–81.

McKay, Chris. *Big Ben: The Great Clock and the Bells at the Palace of Westminster.* Oxford: Oxford University Press, 2010.

McKenzie, D. F. "Printing and Publishing, 1557–1700: Constraints on the London Book Trade." In *The Cambridge History of the Book in Britain*, vol. 4, *1557–1695.* Edited by John Barnard and D. F. McKenzie, 553–67. Cambridge: Cambridge University Press, 2002.

McKitterick, David. *Print, Manuscript and the Search for Order: 1450–1830.* Cambridge: Cambridge University Press, 2003.

McLean, Antonio. *Humanism and the Rise of Science in Tudor England.* New York: Neale Watson, 1972.

McNutt, Jennifer Powell. "Hesitant Steps: Acceptance of the Gregorian Calendar in Eighteenth-Century Geneva." *Church History* 75, no. 3 (2006): 544–64.

Menninger, Karl. *Number Words and Number Symbols: A Cultural History of Numbers.* Translated by Paul Broneer. Cambridge, MA: MIT Press, 1969.

Merton, Robert K. *Science, Technology and Society in Seventeenth-Century England.* Bruges, Belgium: Saint Catherine Press, 1938.

Meskins, Ad. "Mathematics Education in Late Sixteenth-century Antwerp." *Annals of Science* 53, no. 2 (1996): 137–55.

Minns, Chris, and Patrick Wallis. "Rules and Reality: Quantifying the Practice of Apprenticeship in Early Modern England." *Economic History Review* 65, no. 2 (May 2012): 556–79.

Mitchiner, M. B., C. Mortimer, and A. M. Pollard. "Nuremberg and Its Jetons, c. 1475 to 1888: Chemical Compositions of the Alloys." *Numismatic Chronicle* 147 (1987): 114–55.

Mitchiner, M. B., C. Mortimer, and A. M. Pollard. "The Alloys of Continental Copper-Base Jetons (Nuremberg and Medieval France Excepted)." *Numismatic Chronicle* 148 (1988): 117–28.

Mondschein, Ken. *On Time: A History of Western Timekeeping.* Baltimore: Johns Hopkins University Press, 2020.

Muldrew, Craig. *The Economy of Obligation: The Culture of Credit and Social Relations in Early Modern England.* New York: St. Martin's Press, 1998.

Munkhoff, Richelle. "Reckoning Death: Women Searchers and the Bills of Mortality in Early Modern London." In *Rhetorics of Bodily Disease and Health in Medieval and Early Modern England*, edited by Jennifer C. Vaught. London: Routledge, 2010.

Murray, David. *Chapters in the History of Bookkeeping, Accountancy and Commercial Arithmetic.* Glasgow: Jackson, Wylie, 1930.

Neugebauer, Richard. "Mental Handicap in Medieval and Early Modern England: Criteria, Measurement and Care." In *From Idiocy to Mental Deficiency: Historical Perspectives on People with Learning Disabilities.* Edited by Anne Digby and David Wright, 22–43. London: Routledge, 1996.

Newman, William R., and Anthony Grafton, ed. *Secrets of Nature: Astrology and Alchemy in Early Modern Europe.* Cambridge, MA: MIT Press, 2001.

Nobes, C.W. "The Gallerani Account Book of 1305–8." In *Accounting History: Some British Contributions.* Edited by R. H. Parker and B. S. Yamey, 239–49. Oxford: Oxford University Press, 1994.

Noble, Safiya Umoja. *Algorithms of Oppression: How Search Engines Reinforce Racism.* New York: New York University Press, 2018.

Ogilvie, Sheilagh. *Institutions and European Trade: Merchant Guilds, 1000–1800*. Cambridge: Cambridge University Press, 2011.

Ogle, William. "An Inquiry into the Trustworthiness of the Old Bills of Mortality." *Journal of the Royal Statistical Society* 55, no. 3 (September 1892): 437–60.

Okamoto, Yukari. "Mathematical Learning in the USA and Japan: Influences of Language." In *The Oxford Handbook of Numerical Cognition*. Edited by Roi Cohen Kadosh and Ann Dowker, 415–29. Oxford: Oxford University Press, 2015.

Oldroyd, David, and Alisdair Dobie. "Bookkeeping." In *The Routledge Companion to Accounting History*. Edited by John Richard Edwards and Stephen P. Walker, 95–119. New York: Routledge, 2009.

Olson, Richard. *Science and Religion, 1450–1900: From Copernicus to Darwin*. Westport, CT: Greenwood Press, 2004.

O'Neill, Cathy. *Weapons of Math Destruction: How Big Data Increases Inequality and Threatens Democracy*. New York: Crown, 2016.

Ore, Oystein. *Cardano the Gambling Scholar*. Princeton, NJ: Princeton University Press, 1953.

Orme, Nicholas. *Medieval Children*. New Haven, CT: Yale University Press, 2001.

Ormrod, David. *The Rise of Commercial Empires: England and the Netherlands in the Age of Mercantilism, 1650–1770*. Cambridge: Cambridge University Press, 2003.

Ormrod, W. Mark, Bart Lambert, and Jonathan Mackman. *Immigrant England, 1300–1550*. Manchester: Manchester University Press, 2019.

Orser, Charles E. Jr. *An Archaeology of the English Atlantic World, 1600–1700*. Cambridge: Cambridge University Press, 2018.

Otis, Jessica. "'Set Them to the Cyphering Schoole': Reading, Writing, and Arithmetical Education, circa 1540–1700." *Journal of British Studies*, 56, no. 3 (2017): 453–82.

Otis, Jessica Marie. "'Sportes and Pastimes, Done by Number': Mathematical Games in Early Modern England." In *Playthings in Early Modernity: Party Games, Word Games, Mind Games*. Edited by Allison Levy, 131–44. Kalamazoo, MI: Medieval Institute Publications, 2017.

Pagel, Mark, and Andrew Meade. "The Deep History of the Number Words." *Philosophical Transactions of the Royal Society B: Biological Sciences* 373, no. 1740 (February 19, 2018): 20160517.

Parlett, David. *The Oxford Guide to Card Games*. Oxford: Oxford University Press, 1990.

Parry, Glyn. *The Arch Conjuror of England: John Dee*. New Haven, CT: Yale University Press, 2012.

Peacey, Jason. *Print and Public Politics in the English Revolution*. Cambridge: Cambridge University Press, 2013.

Pearson, Robin. *Insuring the Industrial Revolution: Fire Insurance in Great Britain, 1700–1850*. Burlington, VT: Ashgate, 2004.

Perkins, Maureen. *Visions of the Future: Almanacs, Time, and Cultural Change 1775–1870*. Oxford: Clarendon Press, 1996.

Peters, John Durham. "Calendar, Clock, Tower." In *Deus in Machina: Religion, Technology, and Things in Between*. Edited by Jeremy Stolow, 25–42. New York: Fordham University Press, 2013.

Peters, Lien, and Bert De Smedt. "Arithmetic in the Developing Brain: A Review of Brain Imaging Studies." *Developmental Cognitive Neuroscience* 30 (April 1, 2018): 265–79.

Philip, Alexander. *The Calendar: Its History, Structure and Improvement*. Cambridge: Cambridge University Press, 1921.

Piazza, Manuela, Vito De Feo, Stefano Panzeri, and Stanislas Dehaene. "Learning to Focus on Number." *Cognition* 181 (December 1, 2018): 35–45.

Pincus, Steve. " 'Coffee Politicians Does Create': Coffeehouses and Restoration Political Culture." *Journal of Modern History* 67, no. 4 (December 1995): 807–34.

Pinheiro-Chagas, Pedro, Amy Daitch, Josef Parvizi, and Stanislas Dehaene. "Brain Mechanisms of Arithmetic: A Crucial Role for Ventral Temporal Cortex." *Journal of Cognitive Neuroscience* 30, no. 12 (December 2018): 1757–72.

Pinheiro-Chagas, Pedro, Dror Dotan, Manuela Piazza, and Stanislas Dehaene. "Finger Tracking Reveals the Covert Stages of Mental Arithmetic." *Open Mind* 1, no. 1 (January 18, 2017): 30–41.

Plofker, Kim. *Mathematics in India*. Princeton, NJ: Princeton University Press, 2009.

Plumley, N. "The Royal Mathematical School within Christ's Hospital: The Early Years.— Its Aims and Achievements." *Vistas in Astronomy* 20 (1976): 51–9.

Pollnitz, Aysha. "Humanism and the Education of Henry, Prince of Wales." In *Prince Henry Revived: Image and Exemplarity in Early Modern England*. Edited by Timothy Wilks, 22–64. London: Paul Holberton, 2007.

Poole, Kristen, and Owen Williams, eds. *Early Modern Histories of Time: The Periodizations of Sixteenth- and Seventeenth-Century England*. Philadelphia: University of Pennsylvania Press, 2019.

Poole, Robert. " 'Give Us Our Eleven Days!': Calendar Reform in Eighteenth-Century England." *Past and Present* 49 (November 1995): 95–139.

Poole, Robert. *Time's Alteration: Calendar Reform in Early Modern England*. London: UCL Press, 1998.

Poovey, Mary. *A History of the Modern Fact: Problems of Knowledge in the Sciences of Wealth and Society*. Chicago: University of Chicago Press, 1998.

Popper, Nicholas. "An Information State for Elizabethan England." *Journal of Modern History* 90 (September 2018): 503–35.

Porter, Theodore M. *Trust in Numbers: The Pursuit of Objectivity in Science and Public Life*. Princeton, NJ: Princeton University Press, 1995.

Powell, Sarah, and Paul Dingman. "Arithmetic Is the Art of Computation." *The Collation* (blog), September 8, 2015. http://collation.folger.edu/2015/09/arithmetic-is-the-art-of-computation/ .

Prescott, Anne Lake. "Refusing Translation: The Gregorian Calendar and Early Modern English Writers." *Yearbook of English Studies* 36, no. 1, Translation (2006): 1–11.

Principe, Lawrence. *The Aspiring Adept: Robert Boyle and His Alchemical Quest*. Princeton, NJ: Princeton University Press, 1998.

Pullan, J. M. *The History of the Abacus*. London: Hutchinson, 1970.

Quinones, Ricardo J. *The Renaissance Discovery of Time*. Cambridge, MA: Harvard University Press, 1972.

Ramani, Geetha B. and Robert S. Siegler. "How Informal Learning Activities Can Promote Children's Numerical Knowledge." In *The Oxford Handbook of Numerical Cognition*. Edited by Roi Cohen Kadosh and Ann Dowker, 1135–53. Oxford: Oxford University Press, 2015.

Reiss, Timothy J. "Calculating Humans: Mathematics, War, and the Colonial Calculus." In *Arts of Calculation: Quantifying Thought in Early Modern Europe*. Edited by David Glimp and Michelle R. Warren, 137–64. New York: Palgrave Macmillan, 2004.

Relton, Francis Boyer. *An Account of the Fire Insurance Companies*. London: Swan Sonnenschein, 1893.

Richards, Joan L. " 'This Compendious Language': Mathematics in the World of Augustus De Morgan." *Isis* 102 (2011): 506–10.

Richards, R. D. *The Early History of Banking in England*. London: P. S. King and Son, 1929.

Riello, Giorgio. "Things that Shape History: Material Culture and Historical Narratives." In *History and Material Culture: A Student's Guide to Approaching Alternative Sources*. Edited by Karen Harvey. New York: Routledge, Taylor and Francis, 2009.

Roberson, Debi, Jules Davidoff, Ian R. L. Davies, and Laura R. Shapiro. "Color Categories: Evidence for the Cultural Relativity Hypothesis." *Cognitive Psychology* 50, no. 4 (June 1, 2005): 378–411.

Roberts, Gareth, and Fenny Smith, eds. *Robert Recorde: The Life and Times of a Tudor Mathematician*. Cardiff: University of Wales Press, 2012.

Rosenberg, Charles E. "Wood or Trees? Ideas and Actors in the History of Science." *Isis* 79, no. 4 (December 1988): 564–70.

Rossi, Guido. *Insurance in Elizabethan England: The London Code*. Cambridge: Cambridge University Press, 2016.

Rugani, Rosa. "Towards Numerical Cognition's Origin: Insights from Day-Old Domestic Chicks." *Philosophical Transactions of the Royal Society B: Biological Sciences* 373, no. 1740 (February 19, 2018): 20160509.

Russell, G. A. *The 'Arabick' Interest of the Natural Philosophers In Seventeenth-Century England*. Leiden: E. J. Brill, 1994.

Salman, Jeroen. *Populair Drukwerk in de Gouden Eeuw: De Almanak Als Lectuur En Handelswaar*. Zutphen: Walburg Pers, 1999.

Salter, Elisabeth. *Cultural Creativity in the Early English Renaissance: Popular Culture in Town and Country*. New York: Palgrave Macmillan, 2006.

Sarnecka, Barbara W., Meghan C. Goldman, and Emily B. Slusser. "How Counting Leads to Children's First Representations of Exact, Large Numbers." In *The Oxford Handbook of Numerical Cognition*. Edited by Roi Cohen Kadosh and Ann Dowker, 291–309. Oxford: Oxford University Press, 2015.

Saxe, Geoffrey B. "Culture, Language, and Number." In *The Oxford Handbook of Numerical Cognition*. Edited by Roi Cohen Kadosh and Ann Dowker, 367–76. Oxford: Oxford University Press, 2015.

Schotte, Margaret E. *Sailing School: Navigating Science and Skill, 1550–1800*. Baltimore: Johns Hopkins University Press, 2019.

Selin, Helaine, ed. *Mathematics across Culture: The History of Non-Western Mathematics*. Boston: Kluwer Academic, 2000.

Shapin, Steven, and Simon Schaffer. *Leviathan and the Air Pump: Hobbes, Boyle and the Experimental Life*. Princeton, NJ: Princeton University Press, 1985.

Shapin, Steven. "Discipline and Bounding: The History and Sociology of Science as Seen through the Externalism-Internalism Debate." *History of Science* 30 (1992): 333–69.

Shapin, Steven. *The Scientific Revolution*. Chicago: University of Chicago Press, 1996.

Shapin, Steven. *A Social History of Truth: Civility and Science in Seventeenth-Century England*. Chicago: University of Chicago Press, 1994.

Shapiro, Barbara. *A Culture of Fact: England, 1550–1720*. Ithaca, NY: Cornell University Press, 2000.

Shapiro, Barbara. *Probability and Certainty in Seventeenth-Century England: A Study of the Relationships between Natural Science, Religion, History, Law, and Literature*. Princeton, NJ: Princeton University Press, 1983.

Sharpe, Kevin. *Reading Revolutions: The Politics of Reading in Early Modern England*. New Haven, CT: Yale University Press, 2000.

Shenton, Caroline. *The Day Parliament Burned Down*. Oxford: Oxford University Press, 2012.

Shepard, Alexandra. *Accounting for Oneself: Worth, Status, and the Social Order in Early Modern England*. Oxford: Oxford University Press, 2015.

Sherman, Stuart. *Telling Time: Clocks, Diaries, and English Diurnal Form, 1660–1785*. Chicago: University of Chicago Press, 1996.

Sherman, William H. *Used Books: Marking Readers in Renaissance England*. Philadelphia: University of Pennsylvania Press, 2008.

Sherover, Charles M. *The Human Experience of Time: The Development of Its Philosophic Meaning*. New York: New York University Press, 1975.

Sillitoe, Paul, ed. *Local Science vs. Global Science: Approaches to Indigenous Knowledge in International Development*. New York: Berghahn Books, 2007.

Simon, Joan. *Education and Society in Tudor England*. Cambridge: Cambridge University Press, 1966.

Simmons, R. C. "ABCs, Almanacs, Ballads, Chapbooks, Popular Piety and Textbooks." In *The Cambridge History of the Book in Britain*, vol. 4, *1557–1695*. Edited by John Barnard and D. F. McKenzie, 504–13. Cambridge: Cambridge University Press, 2002.

Sinha, Vera Da Silva, Wany Sampaio, and Christopher Sinha. "The Many Ways to Count the World: Counting Terms in Indigenous Languages and Cultures of Rondônia, Brazil." *Brief Encounters* 1, no. 1 (February 24, 2017).

Skeat, T. C. "The Egyptian Calendar under Augustus." *Zeitschrift für Papyrologie und Epigraphik*, Bd. 135 (2001): 153–56.

Skorupski, Peter, HaDi MaBouDi, Hiruni Samadi Galpayage Dona, and Lars Chittka. "Counting Insects." *Philosophical Transactions of the Royal Society B: Biological Sciences* 373, no. 1740 (February 19, 2018): 20160513.

Slack, Paul. "Counting People in Early Modern England: Registers, Registrars, and Political Arithmetic." *English Historical Review* 137, no. 587 (August 2022): 1118–43.

Slack, Paul. "Government and Information in Seventeenth-Century England." *Past and Present* 184 (August 2004): 33–68.

Slack, Paul. "Measuring the National Wealth in Seventeenth-Century England." *Economic History Review*, n.s., 57, no. 4 (November 2004): 607–35.

Slack, Paul. "Perceptions of Plague in Eighteenth-Century Europe." *Economic History Review* (April 2021): 1–19.

Slack, Paul. *The Impact of Plague in Tudor and Stuart England*. Oxford: Clarendon Press, 1990.

Slack, Paul. *The Invention of Improvement: Information and Material Progress in Seventeenth-Century England*. Oxford: Oxford University Press, 2015.

Smail, Daniel Lord. *On Deep History and the Brain*. Berkeley: University of California Press, 2008.

Smith, David Eugene. *Computing Jetons*. New York: American Numismatic Society, 1921.

Smith, David Eugene. *History of Mathematics*. New York: Ginn, 1951–53.

Smith, Eugene, and Louis Charles Karpinski. *The Hindu-Arabic Numerals*. Boston: Ginn, 1911.

Smith, Mark M. *Sensing the Past: Seeing, Hearing, Smelling, Tasting and Touching in History*. Berkeley: University of California Press, 2007.

Smith, Pamela H., and Benjamin Schmidt, eds. *Making Knowledge in Early Modern Europe: Practices, Objects, and Texts, 1400–1800*. Chicago: University of Chicago Press, 2007.

Snyder, Walter F. "When Was the Alexandrian Calendar Established?" *American Journal of Philology* 64, no. 4 (1943): 385–98.

Society of Antiquaries of London. *Proceedings of the Society of Antiquaries of London, 25th November 1909 to 29th June 1911*, 2nd ser., vol. 23. Oxford: Horace Hart, 1911.

Society of Antiquaries of London. *Proceedings of the Society of Antiquaries of London, 27th November 1913 to 25th June 1914*, 2nd ser., vol. 26. Oxford: Horace Hart, 1914.

Stallybrass, Peter, Roger Chartier, John Franklin Mowery, and Heather Wolfe. "Hamlet's Tables and the Technologies of Writing in Renaissance England." *Shakespeare Quarterly* 55, no. 4 (2004): 379–419.

Spufford, Margaret. *Small Books and Pleasant Histories: Popular Fiction and Its Readership in Seventeenth Century England*. London: Methuen, 1981.

Stedall, Jacqueline. *Mathematics Emerging: A Sourcebook 1540–1900*. Oxford: Oxford University Press, 2008.

Steele, Robert, ed. *The Earliest Arithmetics in English*. London: Oxford University Press, 1922.

Stock, Brian. *The Implications of Literacy: Written Language and Models of Interpretation in the Eleventh and Twelfth Centuries*. Princeton, NJ: Princeton University Press, 1983.

Stone, Lawrence. "The Educational Revolution in England, 1560--640." *Past and Present* 28 (July 1964): 41–80.

Stowell, Marion Barber. *Early American Almanacs: The Colonial Weekday Bible*. New York: Burt Franklin, 1977.

Strauss, Gerald. *Luther's House of Learning: Indoctrination of the Young in the German Reformation*. Baltimore: Johns Hopkins University Press, 1978.

Swetz, Frank J. *Capitalism and Arithmetic: The New Math of the 15th Century Including the Full Text of the Treviso Arithmetic of 1478*. Translated by David Eugene Smith. La Salle, IL: Open Court, 1987.

Tannenbaum, Samuel, A. *The Handwriting of the Renaissance: Being the Development and Characteristics of the Script of Shakespeare's Time*. New York: Frederick Ungar, 1967.

Taylor, E. G. R. *The Mathematical Practitioners of Tudor and Stuart England*. Cambridge: Cambridge University Press, 1954.

Taylor, Ed. S., *The History of Playing Cards with Anecdotes of Their Use in Conjuring, Fortune-Telling and Card-Sharping*. London: [Jo]hn Camden Hotten, Piccadilly, 1865.

Taylor, John A. *British Empiricism and Early Political Economy: Gregory King's 1696 Estimates of National Wealth and Population*. Westport, CT: Praeger, 2005.

Thomas, Keith. "The Meaning of Literacy in Early Modern England." In *The Written Word: Literacy in Transition, Wolfson College Lectures 1985*. Edited by Gerd Baumann, 97–131. Oxford: Clarendon Press, 1986.

Thomas, Keith. "Numeracy in Early Modern England: The Prothero Lecture, Read 2 July 1986." *Transactions of the Royal Historical Society*, 5th ser., no. 37 (1987): 103–32.

Thomas, Keith. *Religion and the Decline of Magic: Studies in Popular Beliefs in Sixteenth and Seventeenth Century England*. London: Penguin, 1991.

Thornton, Tim. "Lordship and Sovereignty in the Territories of the English Crown: Sub-kingship and Its Implications, 1300–1600." *Journal of British Studies* 60, no. 4 (October 2021): 848–66.

Thrush, A. D,. and John Ferris, eds. *The House of Commons, 1604–29*. Cambridge: Cambridge University Press, 2010.

Todd, John M. *Luther: A Life*. London: Hamish Hamilton, 1982.

Todd, Margo. "Providence, Chance and the New Science in Early Stuart Cambridge." *Historical Journal* 29, no. 3 (September 1986): 697–711.

Tosney, Nicholas Barry. "Gaming in England, c. 1540–1760." PhD diss., University of York, 2008.

Toulmin, Stephen, and June Goodfield. *The Discovery of Time*. Chicago: University of Chicago Press, 1965.

Towse, John N., Kevin Muldoon, and Victoria Simms. "Figuring Out Children's Number Representations: Lessons from Cross-Cultural Work." In *The Oxford Handbook of Numerical Cognition*. Edited by Roi Cohen Kadosh and Ann Dowker, 402–14. Oxford: Oxford University Press, 2015.

Trenerry, C. F. *The Origin and Early History of Insurance*. London: P. S. King, 1926.

Tribble, Evelyn B. *Margins and Marginality: The Printed Page in Early Modern England*. Charlottesville: University Press of Virginia, 1993.

Tucker, John V. "Data, Computation and the Tudor Knowledge Economy." In *Robert Recorde: The Life and Times of a Tudor Mathematician*. Edited by Gareth Roberts and Fenny Smith, 165–88 (Cardiff: University of Wales Press, 2012).

Tzelgov, Joseph, Dana Ganor-Stern, Aravay Y. Kallai, and Michal Pinhas. "Primitives and Non-primitives of Numerical Representations." In *The Oxford Handbook of Numerical Cognition*. Edited by Roi Cohen Kadosh and Ann Dowker, 45–66. Oxford: Oxford University Press, 2015.

Uittenhove, Kim and Patrick Lemaire. "Numerical Cognition during Cognitive Aging." In *The Oxford Handbook of Numerical Cognition*. Edited by Roi Cohen Kadosh and Ann Dowker, 345–64. Oxford: Oxford University Press, 2015.

Van Egmond, Warren. "The Commercial Revolution and the Beginnings of Western Mathematics." PhD diss., Indiana University, 1976.

Van Egmond, Warren. *Practical Mathematics in the Italian Renaissance: A Catalog of Italian Abbacus Manuscripts and Printed Books to 1600*. Firenze: Instituto e Museo di Storia della Scienza, 1980.

Vanes, Jean. *Education and Apprenticeship in Sixteenth-Century Bristol*. Bristol: Bristol Branch of the Historical Association, 1982.

Varley, Rosemary A., Nicolai J. C. Klessinger, Charles A. J. Romanowski, Michael Siegal, and Dale Purves. "Agrammatic but Numerate." *Proceedings of the National Academy of Sciences of the United States of America* 102, no. 9 (March 1, 2005): 3519–24.

Von Brummelen, Glen Van, and Michael Kinyon, ed. *Mathematics and the Historian's Craft*. New York: Springer Science+Business Media, 2005.

Walford, Cornelius. "Early Bills of Mortality." *Transactions of the Royal Historical Society* 7 (1878): 212–48.

Walker, Jonathan. "Gambling and the Venetian Nobleman c. 1500–1700." *Past and Present* 162 (Feb.ruary1999): 28–69.

Wallis, Patrick. "Apprenticeship and Training in Premodern England." *Journal of Economic History* 68, no. 3 (September 2008): 832–61.

Wallis, Patrick. "Plagues, Morality and the Place of Medicine in Early Modern England." *English Historical Review* 121, no. 490 (2006): 1–24.

Wallis, Patrick, and Cliff Webb. "The Education and Training of Gentry Sons in Early Modern England." *Social History* 36, no. 1 (February 2011): 20–46.

Walsham, Alexandra. *Providence in Early Modern England*. Oxford: Oxford University Press, 1999.

Wardhaugh, Benjamin. "'The Admonition of a Good-Natured Reader': Marks of Use in Georgian Mathematical Textbooks." In *Reading Mathematics in Early Modern Europe: Studies in the Production, Collection, and Use of Mathematical Books*. Edited by Philip Beeley, Yelda Nasifoglu, and Benjamin Wardhaugh, 230–51. New York: Routledge, 2021.

Wardhaugh, Benjamin. "Poor Robin and Merry Andrew: Mathematical Humour in Restoration England." *British Society for the History of Mathematics Bulletin* 22 (2007): 151–9.

Wardhaugh, Benjamin. *Poor Robin's Prophecies: A Curious Almanac, and the Everyday Mathematics of Georgian Britain*. Oxford: Oxford University Press, 2012.

Wardley, Peter, and Pauline White. "The Arithmeticke Project: A Collaborative Research Study of the Diffusion of Hindu-Arabic Numerals." *Family and Community History* 6 (May, 2003): 5–17.

Watt, Tessa. *Cheap Print and Popular Piety, 1550–1640*. Cambridge: Cambridge University Press, 1991.

Wedell, Moritz. "Numbers." In *Handbook of Medieval Culture*, vol. 2. Edited by Albrecht Classen, 1205–60. Berlin: De Gruyter, 2015.

Wells, Robert V., ed. *The Population of the British Colonies in America before 1776: A Survey of Census Data*. Princeton, NJ: Princeton University Press, 1979.

Wernimont, Jacqueline. *Numbered Lives: Life and Death in Quantum Media*. Cambridge, MA: MIT Press, 2019.

Westfall, Richard S. *Science and Religion in Seventeenth-Century England*. New Haven, CT: Yale University Press, 1958.

Whitrow, G. J. *Time in History: The Evolution of Our General Awareness of Time and Temporal Perspective*. Oxford: Oxford University Press, 1988.

Wiese, Heike. "The Co-Evolution of Number Concepts and Counting Words." *Lingua* 117 (2007): 758–72.

Wiese, Heike. *Numbers, Language, and the Human Mind*. Cambridge: Cambridge University Press, 2003.

Wilcox, Donald J. *The Measure of Times Past: Pre-Newtonian Chronologies and the Rhetoric of Relative Time*. Chicago: University of Chicago Press, 1987.

Williams, Burma P., and Richard S. Williams. "Finger Numbers in the Greco-Roman World and the Early Middle Ages." *Isis* 86 (1995): 587–608.

Williams, Jack. *Robert Recorde: Tudor Polymath, Expositor and Practitioner of Computation*. London: Springer, 2011.

Williams, Jack. "The Lives and Works of Robert Recorde." In *Robert Recorde: The Life and Times of a Tudor Mathematician*. Edited by Gareth Roberts and Fenny Smith, 7–24. Cardiff: University of Wales Press, 2012.

Williams, Travis D. "The Earliest English Printed Arithmetic Books." *The Library: The Transactions of the Bibliographical Society*, 7th ser., 13, no. 2 (2012): 164–84.

Willmoth, Francis. *Sir Jonas Moore: Practical Mathematics and Restoration Science*. Woodbridge, Suffolk: Boydell Press, 1993.

Winger, R. M. "Zero and the Calendar." *Scientific Monthly* 43, no. 4 (October 1936): 363–7.

Withington, Phil. *Society in Early Modern England: The Vernacular Origins of Some Powerful Ideas*. Cambridge: Polity Press, 2010.

Woolgar, C. M. *The Senses in Late Medieval England*. New Haven, CT: Yale University Press, 2006.

Wolters, Gezinus, Hanneke van Kempen, and Gert-Jan Wijlhuizen. "Quantification of Small Numbers of Dots: Subitizing or Pattern Recognition?" *American Journal of Psychology* 100, no. 2 (Summer 1987): 225–37.

Wright, Brian. *Insurance Fire Brigades 1680–1929: The Birth of the British Fire Service*. Chalford, England: Tempus, 2008.

Wrightson, Keith. "Popular Senses of Past Time: Dating Events in the North Country, 1615–1631." In *Popular Culture and Political Agency in Early Modern England and Ireland: Essays in Honor of John Walter*. Edited by Michael J. Braddick and Phil Withington. Woodbridge, Suffolk: Boydell and Brewer, Boydell Press, 2017.

Yamey, B. S., H. C. Edey, and Hugh W. Thomson. *Accounting in England and Scotland, 1543–1800: Double Entry in Exposition and Practice*. London: Sweet and Maxwell, 1963.

Yamey, B. S. "Balancing and Closing the Ledger: Italian Practice, 1300–1600." In *Accounting History: Some British Contributions*. Edited by R. H. Parker and B. S. Yamey, 250–67. Oxford: Oxford University Press, 1994.

Yamey, B. S. "The Historical Significance of Double-entry Bookkeeping: Some non-Sombartian Claims." *Accounting, Business and Financial History* 15, no. 1 (2005): 77–88.

Yeldham, Florence A. *The Teaching of Arithmetic through Four Hundred Years (1535–1935)*. London: George G. Harrap, 1936.

Yoon, David. "Counting Tokens from the Excavations at Psalmodi (Gard, France)." *American Journal of Numismatics* 16/17 (2004–5): 173–84.

Zaslavsky, Claudia. *Africa Counts: Number and Pattern in African Culture*. Boston: Prindle, Weber and Schmidt, 1973.

Zerubavel, Eviatar. *The Seven Day Circle: The History and Meaning of the Week*. Chicago: University of Chicago Press, 1989.

Zetterberg, J. Peter. "The Mistaking of 'the Mathematicks' for Magic in Tudor and Stuart England." *Sixteenth Century Journal* 11, no. 1 (Spring 1980): 83–97.

Ziegler, Philip. *The Black Death*. New York: Harper Torchbooks, 1969.

Zupko, Ronald Edward. *British Weights & Measures: A History from Antiquity to the Seventeenth Century*. Madison: University of Wisconsin Press, 1977.

Index

For the benefit of digital users, indexed terms that span two pages (e.g., 52–53) may, on occasion, appear on only one of those pages.